The History of Spacecraft Computers from the V-2 to the Space Station

Patrick H. Stakem

© 2009, 2019

Fifth Edition,
Revised and updated
1st in Space Series

Table of Contents

Introduction

This book documents the development of missile and spacecraft guidance computers from the earliest efforts to the current Space Station and satellite onboard systems. This book developed from the author's presentation at the Johns Hopkins University's Applied Physics Laboratory in 2009 for the Workshop on Flight Software. The second and third editions were expanded with new material and references. The fourth edition updated the material to the state of the art in 2014, with discussions of the latest approaches and architectures, including Orion. More references wee included, and errors were corrected. This sixth edition in 2019, adds new material, and corrects some information. There is coverage of new missions and systems, particularly the emerging Cubesat architecture, and the ISS's spiffy new Supercomputer.

Computers for onboard spacecraft use evolved from the computers used to guide missiles. The computers allowed for a degree of autonomy for the spacecraft, allowing operations to continue without direct ground communications.

The early missile guidance computers were located in underground bunkers, and transmitted their steering commands to the missile via a radio link. The missiles of the day had no inertial guidance (and, GPS was years in the future), and went ballistic after the engine burned out, a period of just several minutes. After that, the laws of physics took over.

The early manned missions such as Project Mercury, were basically a man in a can atop a ballistic missile, and did not incorporate computing power. The later Gemini and Apollo missions relied more and more on onboard compute power, while driving the state of the art.

The early Earth-orbiting spacecraft did not make use of computers, but did have the capability of storing commands onboard for execution at a later time. This storage was sometimes on magnetic tape.

One of the first computers onboard a spacecraft was the OBP (On-Board Processor) on the OAO-C Orbiting (Astronomical Observatory) spacecraft. Later, the Advanced Onboard Processor (AOP) was developed as a follow-on.

This book discusses primarily unmanned US spacecraft, with some discussion of the Saturn and Shuttle computers. Some coverage is included for foreign and planetary missions. It mentions aircraft and missile avionics where applicable. This updated edition includes coverage of the ISS supercomputer, the Orion project, and is current as of the beginning of 2019.

Many thanks for the help I received from many friends and associates in compiling this information.

This book was compiled from ITAR compliant sources.

Author

Mr. Patrick H. Stakem has been fascinated by the space program since the Vanguard launches in 1957. He received a Bachelors degree in Electrical Engineering from Carnegie-Mellon University, and Masters Degrees in Physics and Compute Science from the Johns Hopkins University. At Carnegie, he worked with a group of undergraduate students to re-assemble and get running a surplus Athena missile guidance computer. It was brought up to operational status, and modified for general purpose use.

He began his career in Aerospace with Fairchild Industries on the ATS-6 (Applications Technology Satellite-6), program, a communication satellite that developed much of the technology for the TDRSS (Tracking and Data Relay Satellite System). At Fairchild, Mr. Stakem made the amazing discovery that computers were put onboard the spacecraft. He quickly made himself the expert on their support. He followed the ATS-6 Program through its operational phase, and worked on other projects at NASA's Goddard Space Flight Center including the Hubble Space Telescope, the International Ultraviolet Explorer (IUE), the Solar Maximum Mission (SMM), some of the Landsat missions, and others. He was posted to NASA's Jet Propulsion Laboratory for the MARS-Jupiter-Saturn (MJS-77), which later became the Voyager mission, and is still operating and returning data from outside the solar system at this writing.

His wife likes to remind him that his best software has already re-entered the atmosphere, and burned.

Mr. Stakem is affiliated with the Whiting School of Engineering of the Johns Hopkins University.

Origins of controlled flight

14th century missile guidance

The *Fire Dragon Manual*, by Jiao Yu and Liu Ji, of the Ming Dynasty in China discusses "Arrows of Flaming Fire." At the time, the Chinese were continually bothered by Mongol horsemen, who had managed to conquer most of the continent of Asia, and made head roads into Europe. A major development in weapons technology was required to save the Empire. Although the Chinese were experts at getting things to blow up, it was trickier to get them to blow up when and where they wanted. When the new weapons did blow up, it scared the ponies of the Mongols.

A synopsis of their approach is: "Launch whole bunches of these, and see if we hit something." The advantage of this approach is that no software is required. The disadvantage is a large number of Test & Integration (T&I) fatalities. Things have gone downhill since then.

By the war of 1812, rockets, copied from an Indian design, were in common use for military operation, but had progressed little. The Congreve rockets to bombard Fort McHenry by the British in the War of 1812 were merely a nuisance, but added the line "..rockets red glare..." to the National anthem.

1940's Missile Guidance

Although the Americans, the Russians, and the Germans were experimenting with rockets from the 1920's, the German efforts, spurred on by World War-2 stand out. The V-1, what we would now call a cruise missile, was guided by a simple distance-measuring device driven by a small propeller. It required pre-calculations, what the artillery people call firing tables. You launch from a known position in a known direction. When the distance counter triggered, the missile was put into a steep dive into the target. But the missiles flew low and slow, and many were shot down (some by the Author's father), or were simply flipped out of control by fighter planes. They were not, by any means, an accurate weapon.

The V-2 was a true ballistic missile, developed by the Von Braun team at Peenemunde. Not only did they develop the world's first operational ballistic missile during wartime, but they managed to turn it over to the Army for use in the field. Quite a few were fired against England and the Port of Antwerp with devastating effect. There was no practical defense.

Internal guidance with gyroscopes was used, with the SG-66. (The Smithsonian has one of these). The missiles were launched from a pre-surveyed location in a precise direction. A known distance to the target determined the engine burn time, which was set into the vehicle before launch. After that, no changes were possible. It was inaccurate, but devastating when it worked.

The German V-2 Field Operations Manual was captured by US forces along with missiles and launch and ground support equipment. The manual was translated at the Army's Aberdeen Proving Grounds (MD). It tells the ground troops how to launch the missile. The manual assumes a high school education. After the launch site is accurately surveyed, the missile was erected and fueled. Then, the troops were instructed to "…point fin number one towards London…" The distance was set into the timer that would shut down the engine, the missile was launched, and the support equipment made a hasty withdrawal to avoid Allied air power. This approach was hardly changed as late as the First Iraq War.

Cold War era

After World War-II, the U.S. Government had captured V-2 Missiles and some of the team that developed them, including the leader, Werner von Braun. The U. S. Army Ordnance Corps, with General Electric as a contractor, began assembling and launching V-2's for research, and to get experience with missile technology. The missile developed from the V-2 design was called the Hermes. This program had 5 successful launches, and wrapped up in 1954.

Missile guidance was primitive, relying on programmed sequencers, with limited input from inertial packages. But, this was good enough for a lot of work. In fact, the first stage of the Saturn-V moon rocket used just a simple sequencer for control.

After World War-II, the United States began developing systems to intercept jet aircraft, as it was obvious that anti-aircraft artillery was not going to work. In 1953, the Nike Project, steered to an enemy aircraft by a ground-based controller, was rolled out. The next threat level was an aircraft launching stand-off missiles. Longer range and better intercept capability was needed.

Necessarily, better radars and computational capability were required. These would both be ground-based. The steering commands werq transmitted to the intercept missiles. The initial flight systems were the Nike Ajax, Nike Hercules, and Nike Zeus missiles.

The Bomarc was a surface to air interceptor using rocket engines for launch, and ramjet engines for controlled flight. It had aerodynamic wings, and was controlled in both the boost and cruise modes. When in proximity to the target, the missile was controlled by a onboard seeker. Before that, it was controlled by a ground-based computer, the AN/GPA-35 or AN/FSQ-7. Each ground center could control several missiles simultaneously. Bomarc tests were done from the Air Force Missile Test Center in Florida, and Vandenberg AFB on the West coast.

The Westinghouse AN/GPA-35 was a director for a surface-to-air missile, used with Bomarc. It used a Bendix Radar system, and the ground-based SAGE Computer. Steering commands were transmitted to the flight vehicle. The AN/FSQ-7 Combat Direction Central unit incorporated the largest computer ever built. Each of the 25 units, eventually installed, weighed 250 tons, and were networked. There were a total of 60,000 vacuum tube. Using 3 megawatts of power, the units achieve 75 kilo-instructions per second. Each installation was dual-redundant.

The SAGE computer was ground based. The computer was the AN/FSQ-7 Direction Central Computer, about the same as the IBM 709. A major part of the CPU's time was devoted to making polar-to-rectangular coordinate conversions of the incoming radar data. The program was about 93,000 lines of code, residing on a magnetic drum memory. Each vacuum tube in the AN/FSQ-7 was individually air-conditioned to keep it from overheating

The IBM AN/FSQ-7 Combat Direction Center was a command and control system during the Cold War. It was designed for ground-based interception of incoming enemy bombers, and worked with the SAGE (Semi Automatic Ground Environment) defense network. The core memory was organized as 32-bit words, with a parity bit. It had a 6-microsecond cycle time. The machines were capable of around 75,000 instructions per second. The design came from MIT's Lincoln Labs, with IBM Federal Systems providing the hardware and software. The Automatic Target and Battery Evaluation algorithm used radar data to calculate an interception point for both manned fighter aircraft and Bomarc missiles. The systems could fly the manned system to the target, hands-off for the pilot.

The cpu had an accumulator register, a memory data register, a register for the least significant part of a multiplication, a program counter, and four index registers. The cpu clock was 166 KHz. The real time clock register used a 32 Hz clock. There were no interrupts implemented. Trigonometric functions (sine and cosine) were provided in look-up tables, with binary angles (256 divisions per 360 degrees).

The SAGE system was a derivative of the earlier Whirlwind

computer based systems, developed at MIT. The prototype was completed in 1955, and began running simulated Bomarc interceptions, and live interceptions of target drones flown from the Cape Canaveral station.

The architecture of the Whirlwind computer was precedent-setting, as it was the first real-time system. It was designed at the MIT Servomechanisms Lab, and was influenced by the Eniac design. Construction was begun in 1948. It was a 16-bit design, with 2048 words of memory, using electrostatic storage technology, later changed to the recently developed magnetic core memory.

The architecture of the SAGE computer systems used IBM standard peripherals such as card reader/punches, line printers, magnetic tape and drum memory. The computers were purpose-built to calculate intercepts, and influenced the design of the later Sabre airline reservation system.

The SAGE systems was complex and costly, but never had to launch an intercept for real. In a sense, it helped preserve the peace during 25 years of turbulent times. The concept of linking the control centers with the radar sites over leased telephone lines influenced the later design of the ARPAnet, the predecessor of the Internet. The SAGE machine is preserved in the Computer History Museum in Mountain View, California.

When finally completed in 1963, SAGE had already been made obsolete by nuclear-armed missiles.

The offensive side, ballistic missiles.

Ballistic missiles and launch vehicles need guidance computers as well, for just a few minutes of powered flight. These do not necessarily need to be rad hard, but they have to be ultra-reliable.

The Redstone missile used an inertial guidance unit, and an

onboard guidance computer. The computers were built by the Ford Instrument Company, a division of Sperry Rand. The Redstone was developed by the von Braun V-2 Team, and built by Chrysler Corporation. The first launch was in 1953. A Redstone also launched Alan Shepard on a ballistic trajectory that took him into space in 1961.

The Jupiter missile used an inertial guidance system, and a predetermined trajectory that was followed by the Guidance and Control system. The missile was designed to be launched from a known location, to a selected target location. The guidance and control unit was based on the one for the earlier Redstone missile, and included separate guidance and control computers.

Atlas Computer

The Atlas was the result of the von Neumann ICBM Committee. Atlas A, B, C, and D had no onboard computers and used ground-based guidance Atlas was the first intercontinental ballistic missile deployed by the United States, Weapon System 107A-1, in 1955.

The AN/GSQ-33, Burroughs SM-65 was the ground-based guidance computer for the early Atlas series of rockets. Eventually 17 units were delivered. The ones at the Eastern Test Range (Cape Canaveral) and Western Test Range (Vandenberg AFB) were used for range safety until 1978. The reliability specification was 0.96, but the machine achieved an operational reliability of 0.998. There were no errors during flights. One machine operated 24 hours per day continuously for 17 months without a critical failure. The specified minimum specification for reliability was 0.96. The achieved value was 0.998.

The data word size was 28 bits, and the instruction word was 18 bits. It was a binary machine, with 38 instructions. Clock speed was just above 200 Khz. An ADD operation took 47 microseconds, and a multiply took 532 microseconds.

Memory held 1536 instructions, core had room for 256 items. Some constants could be set by switches. The machine required 208 volt power, 60 cycle, 3 phase, 20 kva from a motor-generator set. Although the cpu used transistor technology, the terminal equipment used tubes. The system included a paper tape reader, paper tape punch, and a plotter.

As a real-time control computer, the machine received missile position and velocity data from radar, and produced steering commands for transmission to the vehicle, during the few minutes of powered flight.

Development of the computer started in 1955. It was a break in tradition from the tube-type computers that were common – it was transistorized. There were three models, developed incrementally. At the time that Burroughs was contracted for the computer, there was no Atlas missile in existence. After the computer was delivered, it was termed the AN/GSQ-33, Digital Computing System. There was some thought of making a truck-mounted "portable" system, but this was never developed.

The machine was implemented with direct-coupled transistor logic. The main problem was the connection between cabinets. If a cable was disconnected while powered, the driver transistor would burn out. This was encountered at the Smithsonian by the author after a large number of failures were noted during debugging of the unit. As with operational installations, the motor generator set was installed on another floor of the building, due to the noise.

The author and another ex-Carnegie graduate with hands-on missile guidance computer experience worked to get a demo program running on the unit. This was not successful, as there were too many transistor failures to keep the machine in a running state. Discussing this with a colleague at work, I found out this was a fault of the transistors available at the time. He had designed the cabinet-cabinet interface. Small world.

14

The Titan-I

The Athena was the ground-based computer for the Titan-I missile. These units were not for flight; they exceeded weight budget by 9 tons. The idea of putting the computer onboard the vehicle was just a dream at this point.

The Univac Athena required 370 square feet of floor space underground in a hardened bunker (some of these survive, and can be toured).. Using radar data input, it calculated course corrections during engine burn. It only had to work for two minutes. It was programmed in assembly language, and was a Harvard architecture, meaning the instructions and data were kept in different stores. In the case of the Athena, the instructions were kept on a magnetic drum, and the data was kept in core.

The Athena cost about $1,800,000. when new, and weighed over 18,000 lbs when shipped. The Athena used a massive motor-generator set with 440-volt 3-phase AC input. The motor generator control unit weighed a ton, and the motor/generator itself weighed over 2 tons.

The machine was built by Sperry Rand Univac in 1957, with Seymour Cray as the chief designer. It had 256 words of 24-bit core memory for data and a 8192-word drum for program and constants. A most interesting feature was Battle Short: In this mode, referred to as "melt-before-fail", the power to the machine could NOT be shut off. Once in service, it was found to have a mean time to failure of 48 days, twenty times better than the original specifications.

The Titan launch complex was located underground, and a single Athena could be used with multiple missiles, launched one at a time. There were eighteen missile complexes in the U. S., each capable of launching multiple missiles. The liquid-fueled Titan's were considered to be only a stop-gap measure pending the deployment of the solid fuel Minuteman Missile, and none of the complexes were operational for more than four years.

The Athena was an extremely reliable computer, which never caused a count-down hold or flight problem in all of those years. The software in the Athena computer operated from a rotating drum, and each instruction had to be spaced along the drum in such a way that the rotation time between instructions was equal to the execution time of the previous instruction.

The last launch supported by an Athena computer was a Thor-Agena missile launched in 1972 from Vandenberg AFB in California. The Athena was used on over 400 missile flights.

The LIM-49 Nike Zeus was an anti-ballistic-missile missile, powered by rockets. It was active in the late 1950's-early 1960's time frame. The ground computer that controlled the flight was derived from the Athena. It used the same 24-bit word size for data, but the instruction size was also increased to 24 bits. Sperry-Rand built the computer, the Target Intercept. The drum memory was gone, and the unit had over 10k of read-only memory for program code and constants, and 2k of read-write memory for variables. There were 5 magnetic tape drives. Much smaller than the Athena, the Target Intercept needed only 70 square feet of space, and weighed 5,200 pounds. In 1961, the memory was upped to a massive 57,344 words. Ten machines were delivered.

A more sophisticated system was required to distinguish real warheads from decoys. This lead to the Nike-X, with a Sperry-Rand CLC (Central Logic and Control) computer. There were multiple processors, dual memory, and dual I/O. It was a 2's complement, 32-bit machine. There was 126,000 locations for program, and 262,000 locations for data. The unit was ready by 1965.

Sperry-Rand also designed small onboard aircraft computers based on the Athena. These were the Univac 1000 and 1020. They used thin film magnetic memory.

The development of onboard missile guidance computers was driven by the need to launch from submarines. Since the launch is not taking place from a pre-surveyed location, the calculations are more difficult. The exact position of the submarine must also be known. This led to the development by the Navy of the Transit system, an early predecessor of the GPS (Global Positioning System). The Transit system relied on a series of Navigation Satellites in known orbits. Once you launch the guidance computer, it only has to work for several minutes, during powered flight, and is not reused.

Titan-II

The Titan-II was used extensively by NASA as a heavy lift vehicle, and was the launch vehicle for the manned Gemini missions. The guidance system was supplied by A. C. Spark Plug, a Division of General Motors, based on an MIT design.

The Titan-II used the IBM ASC-15 guidance computer. The ASC-15 had a drum memory, and occupied a volume of 1 x 1 x 1.5 feet. The drum was 3 inches long, and 4.5 inches in diameter. It had 70 tracks, 58 operational, and 12 spare. Each track held 1,728 bits. The drum held instructions, constants, target data, and timing tracks.

It was later replaced by the Delco Electronics Magic 352, a modified aircraft computer, as part of the Universal Space Guidance System. This system used inertial guidance, and star tracking. The MAGIC 352 had 4k of core memory, using 24 bit words. The cpu was implemented with Fairchild Micrologic integrated circuits. The computer weighed 35 pounds, and occupied a volume of a little over ½ cubic foot. Total power consumption was 90 watts. It was a load-store machine, with a 70 microsecond add, 258 microsecond multiply, and 398 microsecond divide.

Titan-III

The Titan-III is a heavy lift vehicle, and is still in use today. It can lift 7,500 pounds to polar orbit. The ASC-15 flight computer for the Titan-III had a larger drum, with 20 more tracks. This allowed for the storage of 9,792 instructions and 1,152 constants, with a transfer rate of just under 175 kilobits per second. Add time was 156 microseconds, multiply time was 1,875, and division, 7,968. As the Titan-III was used as a space launch vehicle, it got a new computer, the Univac 1824. It was also based on the Athena, in the sense of having 24-bit words. It implemented 2's complement arithmetic, and had thin film memory Instructions were 16 bits, with a 5-bit opcode and 8-bit memory address, and a bit to indicate whether a base register was used in the effective address calculation. It had three index registers. The first unit was delivered in 1968. The 1824, like its sister unit in the P-3 Orion aircraft, used Diode-Transistor Logic, in flatpacks.

The later Titan-IIIc delivered several Viking missions to Mars, and launched dual missions to Jupiter and Saturn in 1977. These spacecraft continued out of the solar system into interstellar space. The Viking landers used a Guidance, Control and Sequencing Computer (GCSC) consisting of two Honeywell HDC 402 24-bit computers with 18K of plated-wire memory. The Viking orbiters used a Command Computer Subsystem (CCS) using two custom-designed 18-bit serial processors.

The 1824 Flight Computer was built in the 1960's by Sperry Rand UNIVAC, Defense Systems Division, a heritage company of Lockheed Martin. It was designed to perform on-board missile guidance and was selected as the guidance computer for the TITAN IIIC missile built by Martin Marietta Corp. Other contracts won by UNIVAC with the 1824 were were the guidance and on-board control of the TITAN IIIC MOL (Manned Orbiting Laboratory) and a maneuverable re-entry vehicle where the computer could guide a re-entry warhead to a ground target. Neither of these programs achieved operational status. The MOL program was scrapped by the Air Force in 1969 partially due to the success in manned flight being achieved by NASA at that time, and the maneuverable re-entry vehicle was never funded by the Air Force for operation.

The 1824 computer was so small that it was taken to the Pentagon for a show-and-tell in a First Class airline seat. Here, Generals were given the opportunity to play what might have been the world's first desk-top video game, a basketball game, using an oscilloscope for a display device. The 1824 was designed in 1963, and production ended in 1969 after undergoing many difficulties, causing it to badly over-run both cost and schedule estimates.

The size was 9" x 9" x 15", with a weight of 32 lbs. The machine used 24 Volts DC power.The Instruction repertoire included 45 unique 16-bit instructions, each comprised of a 5-bit operation code, a 2-bit index register designator, and a one-bit field to indicate whether the extension (base) register was to be used.

A thin-film read/write random access memory provided 512 - 24 bit words. A NDRO (Non-Destructible Readout) Thin-film memory was used for storage of Instructions and constants. There were 3584 48-bit words. Each word contained three 16-bit instructions, or two 24-bit constants.

The computer program was loaded on the ground during pre-launch activities via punched paper or Mylar tape. Magnetic loading media was not allowed by the Air Force due to what they perceived as the vulnerability to magnetic fields. During flight, this on-board computer received vehicle attitude inputs from an Inertial Guidance Platform..

This data was then processed in the 1824, to determine and output steering commands, guiding the vehicle on a predetermined, parameterized path including roll maneuvers, thruster jettisons, and payload release.

The Sabre/1824 was the first Univac defense systems computer to use monolithic integrated circuits. Earlier machines had used discrete semiconductors. These first integrated circuits used Diode-Transistor-Logic (DTL) and were designed by Univac and implemented in silicon by Westinghouse. The 1824 used flat pack integrated circuits. To keep size down, the 1824 used a multi-layer board design.

Minuteman

The Minuteman solid-fuel intercontinental ballistic used an onboard Autonetics D-17 guidance computer. This was based on integrated circuits from Texas Instruments.

The Minuteman-I Autonetics D-17 flight computer used a rotating air bearing magnetic disk holding 2,560 "cold-stored" words in 20 tracks (write heads disabled after program fill) of 24 bits each and one alterable track of 128 words. The time for a D-17 disk revolution was 10 ms. The D-17 also used a number of short loops for faster access of intermediate results storage. The D-17 computational minor cycle was three disk revolutions or 30 ms. During that time all recurring computations were performed. For ground operations the inertial platform was aligned and gyro correction rates updated. During flight, filtered command outputs were sent by each minor cycle to the engine nozzles. Unlike modern computers, which use disks for secondary storage, the disk was the main computer memory. The disk storage was considered hardened to radiation from nearby nuclear explosions. To improve computational speed, the D-17 borrowed an instruction look-ahead feature from the Autonetics's Field Artillery Data Computer (M18 FADAC) that permitted simple instruction execution every word time.

The D-17B and the D-37C guidance and control computers were integral components of the Minuteman-I and Minuteman-II missiles, respectively, which formed a part of the United States ICBM arsenal. The Minuteman-III missiles, which use D-37D computers, completed the 1000 missile deployment of this system. The initial cost of these computers ranged from $139,000 to $250,000.

The D-17B was a circa-1962 synchronous serial machine, with a little over 1500 transistors, and more than 6,000 diodes and discrete components. It weighed 62 pounds, and used about 250 watts of power. It was a binary, fixed point, 2's complement machine, with 39 instructions. The computer and the entire guidance systems was contained in a "guidance ring." that was a part of the structure of the missile.. The machine was a cylinder 20 inches high, with a 29 inch diameter. It is a smaller version of the later Instrument Unit, used on the Saturn vehicles. The machine was programmed in assembly language. It had a mean time between failures of 5.5 years. At the time, the Soviet Union used tube-based missile guidance computers

The D-17C used small scale integrated circuits from Texas Instruments. Thee machine was binary, fixed point, and 2's complement. The add time was 78 microseconds, multiply in 1 millisecond. Division was done in a software routine.

The D-37C used a double-sided, fixed head disk assembly containing 7222 words of 27 bits size (24 data plus 3 spacer bits). The flight software was carefully designed and tested, as the missile carried a nuclear warhead.

The later D-37D used the NS-20 flight computer. This was later replaced by the NS-50, used for the Minuteman-III, which was based on a 16-bit microprocessor.

The machines operated from a 28 volt power supply. They controlled the rocket nozzles, and the stage separation.

The Arithmetic unit held the operand results. It implemented add, subtract, multiply, and divide, as well as logical operations. The input unit supported communications with other units such as sensors and actuators. A program could be loaded from paper tape, or manually from an attached keyboard unit. The power supply was 32 volts dc, and the computer could read the voltage level via a simple A-to-D circuit. The input unit could also receive messages from a radio receiver, in 24 bit chunks. The data rate was 100 bits per second. The output unit drove devices on the missile, such as the gyros. It was also hooked to a transmitter, for telemetry of data to the ground. There were 48 digital inputs, 28 digital outputs, 12 analog, and 3 pulse outputs.

Polaris (UGN-27)

The Polaris missile was a two-stage solid-fuel nuclear-armed submarine-launched ballistic missile (SLBM) built during the Cold War by Lockheed Corporation of California for the United States Navy. It included Inertial guidance and Thrust vectoring for steering.

Size was an issue. A computer small enough to fit through a submarine hatch was developed by 1958, the AN/UYK-1. It was used to interpret the Transit navigation satellite data and send guidance information to the Polaris, which had its own guidance computer made with ultra miniaturized electronics, very advanced for its time. There wasn't much room in a Polaris sub, and there were 16 missiles on each submarine. The details of these computers is still classified. The Ship's Inertial Navigation System (SINS) was developed earlier to provide a continuous dead reckoning update of the submarine's position between position fixes via other methods, such as LORAN. This was important in the first few years of Polaris, because the Transit satellite system was not operational until 1964.

By 1965, microchips similar to the ones Texas Instruments made for the Minuteman II were being purchased by the Navy for the Polaris. The Minuteman guidance systems each required 2000 of these, so the Polaris guidance system probably used a similar number. To keep the price under control, the design was standardized and shared with Westinghouse Electric Company and RCA.

Delta

The Delta launch vehicle has had a long and successful launch history. It was a NASA vehicle, built by McDonnell Douglas, designed specifically for access to Space, not as a ballistic missile. It was based on the earlier Thor missile design. First launch was in 1960, and the rocket is still in use. The Delta vehicle uses a Delco guidance computer, with L3 Communications avionics.

NASA Crewed Missions

This section discusses the Flight Computers in the crewed missions.

Mercury

After the ballistic Mercury missions, the crew, drawn from the military test pilot world, demanded some input in flying the vehicle. The Gemini spacecraft carried a small computer to enable on-orbit rendezvous and docking with the Agena target vehicle, a skill that would be needed for the later Apollo missions to the moon. Gemini, designed for rendezvous, had a computer that could actually take over from the Titan launch vehicle computer. The crew could then fly the vehicle into orbit.

Gemini

The Gemini Guidance computer, built by IBM Federal Systems Division's Space Guidance Center in Owego, N.Y, weighted 59 pounds. It had 16 machine language instructions, with a 140 millisecond ADD time. Equations were verified at the Fortran level. Auxiliary tape memory was used (on Gemini VIII and later) and glass delay line registers were used for temporary storage.

An IBM guidance computer was used on all Gemini flights, including the first rendezvous mission, between Gemini 6 (Astronauts Walter Schirra and Thomas Stafford) and Gemini 7 (Astronauts Frank Borman and James Lovell). Occupying only 1.35 cubic feet of space, the onboard system performed some 7,000 calculations a second to bring the two vehicles nose-to-nose, 120-feet apart, 185 miles above Hawaii. The computer had a memory system capable of holding nearly 20,000 bytes of information.

During a Gemini flight, five IBM S/360 mainframe computers in

Houston, Texas, performed 25 billion calculations every 24 hours to provide NASA flight controllers with almost instantaneous reports on the moment-by-moment progress of the mission.

The onboard computer had an average power consumption of 94.54 watts, a 500 kHz bit rate, a memory cycle time of 250 kHz and an add time of 140 microseconds. The computer's memory was a random-access, nondestructive readout design with flexible instruction and data storage organization. Its nominal capacity was 4,096 39-bit words and its operational capacity was 12,288 13-bit words.

The computer accepted data from gyros and similar onboard systems, from the astronauts, and from ground control systems, then computed and reported navigation and control information to the two astronauts. During ascent and reentry, the IBM computer could also issue steering signals to control the flight path.

Apollo/Saturn

The flight computers for the Saturn launch vehicle were an evolution of earlier missile guidance efforts. The massive Saturn-V first stage was built from clustered Jupiter rockets. The earlier and smaller Saturn-I was built of clustered Redstone rockets, which were themselves a derivative of the early German V-2 rockets by the Von Braun Team. The Saturn's upper stages were all new technology, using liquid hydrogen and oxygen. The vehicle had to achieve and maintain a precise trajectory from its launch site in Florida, to lunar orbit, to the lunar surface and back, and then return to Earth. All this took unprecedented computing power.

The Saturn-1 launch vehicle used the same control computer as the Titan, the ASC-15. A magnetic drum of less than 100k words capacity, was used as memory. It was a simplex design, with no redundancy. It weighed under 100 pounds, and consumed 275 watts of power.

A sequencer design from the Jupiter missile was used for timed events. This was a 6-track tape recorder that drove relays to activate and deactivate components at the right time in the flight. Guidance was open-loop, with commands only as a function of time.

In 1964, the 99-pound IBM computer became the first onboard computer into space. This was as a passenger, doing the computations but not steering the vehicle.

Onboard the Saturn SA-6 vehicle, the IBM system handled the thousands of complex calculations needed to determine the position and velocity during launch, it was capable of compensating for an emergency engine shutdown by issuing a command to redirect the thrust of other engines. It replaced a mechanical cam that implemented the tilt program on earlier vehicles.

The Saturn Launch Vehicle Guidance Computer was used on the Saturn IB and Saturn V rockets. It has a 26-bit word size, with two additional parity bits. Instructions were 13 bits long. The memory was core, augmented with ultrasonic delay lines for temporary storage. Triple Modular redundancy was used. The computer weighed about 72 pounds, and used 137 watts of power. It only had to operate for a short period of time, guiding the 3,000 ton Saturn-V to 100 miles altitude, and into the proper orbit.

The clock was 2.048 MHz. Memory size was 32k of 28 bit words. The SLVGC was successful in getting all of the vehicles to orbit, and the ones destined to the Moon on the proper trajectory.

The giant Saturn-V first stage used sequencing done by a magnetic tape, a conservative yet reliable approach. Actually, the astronauts could take over control of this launch vehicle from the capsule, but as far as I know, this never happened.

Saturn-V first stage, S1C

The first stage of the Saturn Rocket "stack" was the heavy lift stage, consisting of five Rocketdyne F-1 engines, one fixed in the middle, and four outside units that could swivel for steering and attitude adjustment. The first stage booster did not incorporate active guidance. The stage's job was to get the rocket and its payload from a standing start to 67 kilometers up, 93 kilometers downrange, and moving at 2,300 meters per second. That required 168 seconds of engine burn time. The total thrust developed by the engines was 7,600,000 pounds-force. Most of the first stage was fuel. The dry weight was about 130 tons, and the fueled weight was 2,300 tons. Any deviation of the vehicle during first stage burn was noted, and adjusted for during the second stage burn.

The engine's sequence of events was controlled by an onboard sequencer. This was not a computer, but just a fixed series of commands that were played out in time sequence. The center engine of the stage was started 8.9 seconds before launch, with pairs of outboard engines starting at 300 millisecond intervals. This technique was used to reduce structural loading on the rocket. When the computer in the Instrument Unit confirmed thrust level correctness, the pad hold-down arms released the rocket. In the Instrument Unit, the Saturn Emergency Detection System (EDS) inhibited engine shutdown for 30 seconds after launch. It was calculated that this was safer than having a shutdown early in the sequence, which would result in a non-survivable event for the astronauts.

The Saturn V second stage, S-II, and third stage S-IVB were controlled by the computer in the Instrument Unit. About 38 seconds after second stage ignition, the vehicle control switched from a preprogrammed trajectory to closed loop control. The Instrument Unit computed in real time the most fuel-efficient trajectory toward its target orbit. If the Instrument Unit failed, the crew could switch control of the Saturn to the Command Module's computer, take manual control, or abort the flight.

The Instrument Unit (IU) was the center point of the data flow on the Saturn vehicle, sending and receiving data both up and down the vehicle. It was introduced with the Saturn-I, Block–II, unit SA-5. It was designed at MSFC, and produced by IBM Corp. It used for vehicle guidance, control, and sequencing. The IU had its own telemetry, tracking, and power systems. The first model was a ring structure 154 inches in diameter, and 58 inches high. The large diameter matched the profile of the launch vehicle. Version two of the Instrument Unit was 34 inches high and 21 feet in diameter. It was constructed of an aluminum honeycomb, less than an inch thick, and weighed 2,670 pounds. It was unpressurized, unlike the previous version. Subsequent Saturn-V vehicles used a third version. The IU was placed between the S-IVB second stage and the Apollo spacecraft payload, and was an integral structural member. The IU was cooled by a water/methanol heat exchanger, and powered by batteries.

In the Saturn-V configuration, the IU Launch Vehicle Digital Computer (LVDC) had 6 memory modules of 4096 28-bit words. The computer could achieve 9,600 fixed-point operations per second. An Add or Subtract operation took 82 microseconds, and Multiply and Divide, 328. Add/Subtract, and multiply/divide could take place simultaneously. There were twelve prioritized interrupts. All hardware was triplicated, for reliability. The computer had an associated Launch Vehicle Data Adapter, handling input and output. Besides attitude correction calculations, the digital computer handled sequencing of events, such as staging. Memory was based on ferrite cores.

In addition to the digital computer, the IU had an analog Flight Control Computer. The digital unit monitored status and calculated attitude corrections. The analog computer was used to command these corrections as angular adjustments for the swiveling nozzles.

The unit was powered by four 28-volt batteries. Heat generation in the unit was handled by a liquid cooling loop, which dumped heat to the outside air. Before launch, a precision theodolite was used to align the inertial platform in the IU to the exact launch azimuth. The ST-124 inertial platform switched from an Earth Reference to a space-based frame of reference 5 seconds before liftoff.

The ST-124 3-degrees of freedom Inertial Platform Assembly, produced by Bendix Corporation, were a derivative of similar gyro-based platforms used on the German V-2 missile. It included three single degree of freedom precision gyros, and accelerometers. Active guidance was not required during the main boost phase, as the pre-programmed first stage's job was to just get the vehicle up and moving. Any needed adjustments would be made during the second stage burn.

The IU included a data adapter to interface with the various sensors and other systems. This unit transformed the various formats, including analog, to a standard digital format the digital computer could use.

Before launch, the LVDC was connected to the ground control computer via umbilical.

Control of the first stage was based only on a time sequence. True guidance was not applied until after the second stage burn had been initiated. Engine cut-off was determined by having achieved a velocity sufficient to enter Earth orbit. The algorithm was a minimum propellant flight path, using calculus of variations. For the second and subsequent stages, closed loop control was used.

When the IU sensed that the propellant level in the first stage had reached a preprogrammed value, it commanded stage cut-off and the initial separation sequence. It handled a similar role for the second stage. The IU controlled the first burn of the third stage, to achieve Earth orbit, and the second burn, to enter the trans-lunar trajectory. The IU had done its job at this point, about 6 ½ hours after launch.

Until a large enough aircraft became available, the IU was delivered from Huntsville to the Cape via barge. In one case, schedule pressure required construction of a portable clean room on the barge, and work on the IU was completed en route.

A flight spare IU can be seen at the National Air and Space Museum, Steven F. Udvar-Hazy Center, near Dulles Airport in Virginia.

The LVDC in the Instrument Unit was developed by IBM Corporation. It was used on the Saturn I Block-II and Saturn-V. The computer used inputs from the ST-124 inertial platform to calculate trajectory and navigation. The processor was serial, using fixed-point data of a 28-bit word size. The hardware was built up from modules containing discrete components. There were two arithmetic logic units (ALU's), one for addition, subtraction, and logical operations, and the other for multiply and divide. These could operate independently. The CPU included the two ALU units, and a set of registers to hold data. The serial registers were implemented as ultrasonic glass delay lines.

The computer clock was 2.048 MHz, and an instruction cycle took 82 microseconds. Most instructions used a single cycle, but the longest one took 5 cycles. The memory was 32 kilobytes of 28-bit words. Within these 28 bits were 26-bit sign-magnitude data words, with two bits of error detection. Instructions were 13 bits long, with parity. Two instructions were packed into a memory word. Instructions had a 4-bit operation code, and a 9-bit address field. There were 18 instructions in the instruction set. Add time was 82 microseconds, multiply 328, and divide 656. The complexity of the unit was equivalent to over 40,000 transistors. It could achieve an instruction execution rate of just over 12,000 instructions per second. The computer weighed 75 pounds, and required 150 watts of power. There was some early discussion of replacing the AGC with an LVDC, but this was not pursued.

Memory used magnetic core technology, with delay lines for

temporary storage. It featured serial read. A maximum of eight modules of 4096 bits each could be used.

Upon RESET, the CPU fetched the starting address from memory address zero. This was used as an operand for an unconditional jump ("HOP") instruction. There were specific I/O instructions. The machine was programmed in assembly language. No source code examples still exist. There is some evidence that the program was originally produced by engineers in Fortran, which would make sense, because it could then be run as a simulation. The Fortran was then hand-assembled into LVDC code.

The circuitry was triple-modularly-redundant (TMR) with voting (using disagreement registers), and, in some critical cases, quadruple redundant. The triple seven-stage instruction execution pipelines had voting circuitry at each stage. The published reliability number was 99.6% over 250 hours of operation, although the unit was only required to operate for minutes in the early launches. There were 11 hardware interrupts on the early Saturn I-B models, including one for engine cut-off. The LVDA unit also served as the interrupt controller, receiving all the interrupts from external devices, and signaling the CPU over a single line. Interrupts could be masked (ignored).

The digital computer had an associated Launch Vehicle Data Adapter (LVDA), which was an I/O interface to the inertial platform, the command receiver and telemetry transmitters, the RCA-110 ground checkout computer (while on the launch pad), and other vehicle sensors, such as separation switches. The LVDA also provided analog-to-digital conversion. It communicated with the LVDC over a 512 kbps serial interface. The LVDA input link from the RCA–110 ground computer was a 14-bit data line. The LVDA weighed 214 pounds, and required 320 watts of power. Local storage of data in the LVDA was via glass delay lines The LVDC and the associated LVDA was a real-time control computer. Previous versions of the Saturn had used open loop control.

The LVDC computer system also did pre-launch self-test and supported mission simulation. Its primary purpose was booster guidance. In mission SA-6, one engine shut down prematurely, but the computer automatically adjusted the trajectory to compensate properly. 28 vdc power was supplied from alkaline silver-zinc battery packs

The Flight Control computer in the IU was an analog machine. It took computed attitude correction commands from the LDVC and attitude state data from the ST-124-M It commanded the roll control for the first stage, and pitch, roll, and yaw control for Stages 2 and 3.

AGC

The iconic Apollo Guidance Computer (AGC) was developed by the MIT Instrumentation Lab, headed by Charles Stark Draper, based on the Polaris submarine-launched missile guidance computers. The Project kicked off in 1961 with a 1-page specification from NASA. The unit was designed at MIT and built by Raytheon Missiles and Space Division. This was an overwhelmingly difficult task, given the state of the technology at the time. The integrated circuit had been invented only 2 years earlier. The early model computer used a core-transistor logic, and later models used a single type of NOR integrated circuit after October 1962. A prototype was built in 4 racks the size of modern refrigerators. Using a single type of integrated circuit building block simplified the procurement of parts, which were supplied by Fairchild and Signetics. An approach for in-flight repair by the astronauts for the guidance computer en-route to the moon was considered. A soldering iron was to be included in the Apollo capsule. Later, the reliability of the unit precluded this approach. The tricky part was re-packaging the four floor to ceiling racks of circuitry into a small box. A simple matter of detail.

Before an actual AGC had been implemented, simulations were

run on the Lab's Honeywell 1800, or IBM S/360 mainframe computers. The simulations ran about 10 times slower than real time.

Since a single type of logic gate was used, these could be combined on a logic board that held 60 circuits. These boards were also used in the construction of simulators and test equipment for the computers. Initial delivery of a computer to MIT took place in August of 1964. Ten more units were under construction at Raytheon. They were delivered late, but the spacecraft was behind schedule as well.

The later model of the AGC added a divide and a subtract instruction. The AGC was critical for guidance and navigation. The computer was a 16-bit, 1's complement machine, with a 1.7 microsecond, 12-step cycle time (current machines are sub-nano-second). It had 2048 bytes of random access memory, and 36k of read-only memory, both implemented in a magnetic core technology. The software was released in January of 1966, with the first flight was in August 1966. The design was used until 1975. No in-flight errors were ever attributed to software. None. This was after 2,000 person-years of independent verification and validation (IV&V).

The read-only memory, also implemented in magnetic core technology, was referred to as "core ropes." Here, the default was storing a "1", and if a "0" was desired, a wire was not passed through the core. They were linear structures, not 2 or 3 dimensional. When the coders had done their job, a map of 1's and 0's was passed on to Raytheon, who manufactured the core ropes, which generally changed with each mission. Only ladies well practiced in knitting had the patience to hand-manufacturer these. Our dominance in the space race with the Russians depended on the dexterity of US needle workers.

The later model AGC had 11 instructions, including a double precision multiply. The unit included 20 counters, 60 discrete inputs, 18 discrete outputs, and 5 interrupts. Parity calculations were applied to items stored in memory.

The software for the AGC contained an Executive program, that arranged the order of execution for up to 7 independent program modules. A priority scheduling scheme was implemented by a program called Waitlist. In lieu of a modern real-time operating system, the Executive scheduled tasks in priority order. It could handle 8 tasks, one executing, and 7 waiting. In 1968, there were about 400 programmers working on the AGC code.

Telemetry from the AGC used a single telemetry channel operating at 64 Khz. In this system, a Group was 1 second of data, 64k bits. A Frame was 20 mSec of pulses, or 1280 bits. A group contained 50 frames. A telemetry word was 8 bits, and 160 words fit in a frame.

The AGC had an associated display and keyboard unit for human interaction. This was called the DSKY. The keyboard had 16 keys. The display was electroluminescent, seven-segment format, and displayed 18 decimal digits and 3 signs. In actual use, an Apollo mission would require more than ten thousand keystrokes by an Astronaut.

The power supply for the AGC supplied 13 volts at 2.5 amps and two 3-volt lines at 3 amps and 22 amps. The AGC could be placed in a reduced power, stand-by mode.

The AGC had failure-detection built in. An alarm light was activated by a fault, and the machine continued to run. The overhead cost of the error checking circuitry was estimated to be 5%. Errors could be caused by a failed parity check, clock failure, a trap instruction, a timer on the interrupted state, or a power failure.

There were two Apollo Guidance computers per mission, one in the Command Module; and one in the Lunar Lander. This proved to be a good idea on Apollo 13, which suffered an explosion that crippled the Command module power system on the way to the moon. The computer in the Lunar Lander was re-tasked to provide guidance computations to get the astronauts back to Earth, before the Command Module would be re-activated for re-entry.

The guidance computers had 152 kilobytes of storage for the entire mission. The size was 6 inches, x 1 foot x 2 feet; they weighed 70 pounds, and used 55 watts of electricity. They were constructed of 5600 3-input nor gates, and featured a cycle time 11.7 microseconds.

RTL logic gates were from Fairchild Semiconductor, about 60% of the total US production of microcircuits at the time. The computer was a 16-bit machine, and had a 1.7 microsecond cycle time (current machines are sub-nano-second). It had 2048 bytes of random access memory, and 36k of read-only memory, both implemented in a magnetic core technology. There were four registers, the accumulator, the program counter, the remainder from the DV instruction or the return address after a transfer of control instruction, and the lower product after a multiply instruction. There were five vectored interrupts. The clock was 1.024 MHz.

MIT's instrumentation lab had an early IBM model 650 mainframe for development and test, and rented time on 704's, 709's, and 7090's when needed. The IBM 650 got replaced by a Honeywell H-800. This was a 48-bit machine produced in conjunction with Raytheon. An instruction included an opcode and 3 operands per word. Four banks of core memory, each 2048 words, could be added. It had up to 4 magnetic tape drives, a printer, and a card reader-punch. It's assembly language was called Argus, Automatic Routine Generating and Updating System. It could run the Fortran language. The machine could barely keep up with the increasing demands, and was upgraded to an model 1800 by 1964. Up to 12 more memory banks could be added, and 8 megabytes of disk secondary storage.

A hybrid simulation machine based on Beckman analog computers and a SDS-9300 digital computer were used for Command module and lunar module cockpit simulations.

Although the Saturn Booster used its own guidance system, the AGC monitored launch parameters and provided indications that the launch was on course. The program was started automatically or by astronaut input.

The AGC interfaced with the Inertial Reference Platform in the Command Module, and with the astronauts, via a keypad and numeric display.

The calculations were done internally in metric, but the astronauts (mostly test pilots) preferred English units for display. (What's the worst that could happen?)

An HP-65 hand held scientific calculator was carried on the later Apollo-Soyuz mission, circa 1975, to perform calculations for the rendezvous maneuvers. This was a backup to the Apollo computer onboard the craft. By that time, the complexity of the HP-65 exceeded that of the AGC.

The Lunar Module also had a separate Abort Guidance System computer (AGSC). If the primary guidance system (which included the AGC) failed, the AGSC could be used to return the LEM to a safe lunar orbit, and await rendezvous by the Command Module. It was not capable of accomplishing a lunar landing. It was custom-built by TRW. The computer was called the MARCO 4418, MARCO referring to "MAn Rated COmputer". It weighed about 32 pounds, and used 90 watts of power. The memory was 18-bit serial access. There was 4096 words of memory, the lower half being RAM, and the upper half being ROM. Integer data types were 2's complement, and addresses were 13 bits. The machine had several registers, including an accumulator and an Index Register. There were 27 different instructions, and the software was written in assembly language. This was a real-time control computer, operating different tasks on a major cycle of 2 seconds, with minor cycles of 20 or 40 milliseconds. It was never required, but was used as a backup on Apollo 11.

Skylab and 4Pi

By the time that the Saturn rocket had put men on the moon and returned them safely multiple times, the Apollo computers were mostly obsolete. This was due to major advances in hardware and software, largely driven by the Apollo effort. With some spare Saturn-V's sitting around, the next project was the Skylab space station. This used a Saturn IVB stage as the structure for the station, launched by a Saturn-V with live first and second stages. Astronauts were carried to the facility already in-orbit on three missions in 1973-1974 by Apollo capsules on Saturn-Ib vehicles. Skylab was in orbit until 1979, when it reentered the atmosphere.

With a large volume of science data to be managed, the AGC was not up to the job. The IBM System/4Pi TC-1, a derivative of the IBM S/360 mainframe, and a relative of the subsequent Shuttle's AP-101, was used. This was a radiation-hardened unit, with a 16-bit word, and 16k of memory. It had a custom-designed input-output unit for the lab. It drew 56 watts of electrical power, and weighed 18 pounds. It was built with ttl-technology integrated circuits, and core memory. Ten were built, and two were flown. A standard S/360 mainframe was used to produce code for the 4Pi, and a simulator was used for verification. An IBM System/360-75 was used to evaluate the onboard computer's performance in orbit. It could run a simulation at 3.5 times less than real-time. There was also a hybrid simulator, with a System/360 model 44 hooked to a 4Pi.

The 4Pi had 54 different instruction, and supported 32-bit double precision data. Cycle time was 3 microseconds. An add took 3 cycles, and a multiply or divide took 16. It had a 24-bit real-time clock. A simple numeric keyboard was used to communicate with the computer in the lab in orbit, and it could also be loaded via uplink, and by an onboard tape drive.

The first Skylab mission lasted 272 days, followed by an unmanned period of 394 days, when the computer kept things going. The computer was turned off for 4 years while NASA discussed reboosting Skylab to a higher orbit, or letting it reenter. There was a need to put some mods in the software, but the tools and card decks containing the code had been discarded. This resulted in some 2500 cards being re-punched from listings. At the end of 4 years, the onboard computer was booted up by ground command, and the updates worked fine.

Shuttle

This section discusses the flight computers for the Space Shuttle. These controlled the crewed vehicle in a variety of modes, including launch vehicle, space vehicle, and as a winged, unpowered aircraft, landing on a runway.

The Space Shuttles carried five identical computers, the circa-1972 AP-101's, derived from the IBM System/360 System /4 Pi mainframe architecture. It was a 32-bit machine with 16 registers, and was microprogrammed. It had an instruction set of 154 opcodes. One of the five AP-101's on the Shuttle contained software derived independently from the software loaded on the other four. Each unit had a CPU and an IOP - Input/Output Processor. Each IOP had 24 channels, each with its own bus and processor. Triple redundant power supplies, fed by separate essential electrical buses on the Shuttle were used. The computers were located in three separate locations in the Shuttle Orbiter.

The AP-101 computer was also used in the USAF's F-15, B1B, and B-52 aircraft.. The original model used discrete TTL logic and core memory. The 1990's AP-101S upgrade used semiconductor memory with battery back-up. The Shuttle computers were programmed in the HAL/S language. The USAF used Jovial. Main memory size was 81,000 words.

An Odetics Mass memory unit was added later, with a capacity of 8 million 16-bit words. Different software loads were used during different operational modes such as ascent, on-orbit, and reentry. Later additions to the Shuttle flight deck included a "glass cockpit" with displays for the pilot and mission commander. There were no in-flight failures of the Shuttle computer system. None.

Initially, the Shuttle was not launched if its flight would roll-over from December to January. Its flight software, designed in the 1970s, was not designed to handle this, and would require the orbiter's computers be reset through a change of year, which could cause a glitch while in orbit. In 2007, NASA engineers devised a solution so Shuttle flights could cross the year-end boundary.

ISS

In 1993, the United States and Russia joined to create a International Space Station. On-orbit construction began in 1998, and was completed with a last Shuttle mission in 2011. It is the largest artificial satellite in Earth orbit, and can be seen from the ground with the naked eye. The ISS is a synthesis of several space station projects from America, the Soviets/Russians, the Europeans, and the Japanese.

The Station has provided and uninterrupted human presence in space since 31 October 2000. It is operated as a joint project between the five participanting space agencies.

The International Space Station is continuously crewed, and orbits the Earth at an altitude of some 250 miles. It is quick, traveling at 17,300 miles per hour. It is also expensive, representing an investment of some $100+ billion dollars by the world community, mostly by the United States and Russia. It is thus the most expensive object ever constructed by mankind. It has been visited by astronauts and cosmonauts from some 15 nations, and by paying tourists.

The onboard computing infrastructure was based on rad-hard Intel 80386's for housekeeping, environmental monitoring and control, and station-keeping (orbital maneuvers).

There is an Ethernet backbone in the Station, as well as WiFi. Applications supported include IP phone with webcam for crew conversations with family's.

The laptops replaced the original design of the Multi-Purpose Application consoles, which were MIL-Spec equipment, purpose-built and programmed from scratch using the ADA language and X-windows. The laptops (PGSC – Portable General Support Computers) were more flexible. The general concept is that COTS machines can display any data, but must follow a arm/check/fire protocol to send commands. Originally, the support computers were Grid laptops. These were 8086-based machines, running GridOS.

The Station used some 68 COTS IBM/Lenovo ThinkPad A31 laptops, 32 ThinkPad T61 laptops, and a T61 as a server, with routers. The laptops handle non-critical and experimental support as well as music and television for crew entertainment. The computers run Windows-XP, with some 3 million lines of flight code. The architecture was designed as a station-wide distributed system.

The IBM/Lenovo laptops were modified to handle the environment of the station. With no convention cooling, due to lack of gravity, fans had to be added to keep components on the motherboard cool.

In addition, the circuit boards and connectors were conformally coated to contain any debris. The power supplies were modified to accept Station 28 vdc power. In addition, the laptop had to operate in the 10 psia atmosphere of the Space Shuttle, even though the Station maintains a sealevel pressure of 14.7 psia. The laptops were also modified to have Velcro for attaching to convenient surfaces.

Just like any other pc, the Station laptops were vulnerable to computer viruses and hackers. In August of 2008, some of the onboard laptops were infected with a worm used to steal information from systems.

Initially running the Windows operating system, the laptops were transitioned to Linux. The laptops were also upgraded to HP Z-books. These are modified commercial units, that use less power than their Earth-bound family. These use Intel Xeon cpu's, and come in at 4.4 lbs. This is good when it costs $10,000 per pound for shipping.

The latest addition to the ISS computer network is a pair of supercomputers, They went up in a Dragon capsule, launched by a Falcon-9. It is a COTS Hewlett Packard Enterprise Apollo, with high speed interconnect. It runs the Linux operating system, an achieves a performance of 1 teraflop (10^{12} floating point operations per second). A twin is kept running on the ground.
It is the pizza box server form factor. There are two racks of servers, and include X86 architecture cpu's with solid state storage drives. It is water-cooled, and the power supply was modified to use station power. The running software will monitor the computer's operation to check for the symptoms of radiation damage, which normally manifests itself as an increase in current draw. This can sometimes be overcome with a reboot.

The machine has several different purposes. First, it will show if a computer in space, normally subject to radiation induced upsets, can survive within the crewed space of the ISS and operate for a year. That is about the limit for crews. The second reason for the installation is to "take the computer to the data." It will crunch experiment data and allow downlinking of processed information instead of the raw data, vastly increasing the required bandwidth.

In addition, HP announced that NASA is going to Phase-in second generations laptops on the station, the Zbook 15 mobile workstation. These will replace the HP Z-series 8570's now in use. There are currently 100 laptops and workstations onboard.

The machine was made by Hewlett-Packard, and is a COTS unit. For everything it does, it is "shadowed" by an identical system in an HP facility in Wisconsin. The design is pizza-box form factor, for HP's Apollo40 family, with Intel multi-core Broadwell processors, and a 56 Gigabit per second interconnect. It is rated at a Teraflop, and runs the linux operating system. HP says, "take the computer to the data.
For everything it does, it is "shadowed" by an identical system in an HP facility in Wisconsin. The design is pizza-box form factor, for HP's Apollo40 family, with Intel multi-core Broadwell processors, and a 56 Gigabit per second interconnect. It is rated at a Teraflop, and runs the linux operating system.

The Space Station hosts over 100 Mil-STD-1553 buses (16 bit, 1 Mbps), and GPS/GLONASS navigation systems. Sensors include Sun and star sensors and gyros.

The communication links to and from orbit use Ku band communications, relayed through the Tracking and Data Relay Satellites, providing 3 Mbps up/10Mbps down, and continuous coverage. Email is synced from the ground every 8 hours.

In nine years of operation, there were only two cases of requiring a power off/power on cycle to fix a problem. The Station receives 24x7x365 support from the Johnson Space Flight Center in Houston. There is a duplicate data system on the ground.

Constellation Program

The Constellation Program was NASA's follow-on to the Space shuttle, and was to provide a means for crewed spacecraft to return to the Moon. The stated goals of the program were to gain significant experience in operating away from Earth's environment, develop technologies needed for opening the space frontier, and conduct fundamental science. Necessarily, the onboard computer systems would be more sophisticated than those of previous vehicles that have flown in space. The program included the development of both new booster rocket systems, and new spacecraft.

Modules included the Altair, or Lunar Surface Access Module which drew heavily on the previous Apollo lunar lander design. The computer system was expected to be identical to that used in the Orion. The program was canceled in 2010, but parts survive.

Orion

Orion is the name of the Crew Exploration Vehicle, currently being built by a Lockheed-Martin Team. Team member Honeywell is supplying the computers. The three flight control modules are derived from the Boeing 787 avionics project. This architecture consists of multiple computer systems linked in a network. The initial missions of Orion will be flights to the International Space Station for crew substitution and resupply of consumables. The Orion capsule will be reusable, and supports a crew of four. A first, unmanned launch was successful in 2014 by a Delta-IV Heavy vehicle out of Launch Complex 37 at Cape Canaveral. The capsule was recovered at sea. A flight to dock with the International Space Station is planned.

The Orion avionics, including the computer, is adapted from an advanced aircraft computer. It has a networking architecture, has multiple redundant units, and is radiation-hard. The Vehicle

Management Computer is an adaption of Honeywell's Flight Computer for the 787 aircraft. This is based on the IBM PowerPC 750FX. This unit was introduced in 2002, and has an on-chip level 2 cache of 512 kbytes. Some of the cache can be locked. For radiation hardness, it is manufactured in silicon-on-insulator technology. It has 39 million transistors. It consumes 4 watts or power, operating at 800 MHz. In commercial versions, it was used on the Apple iBook G3. Each Flight Computer has 2 processors. The avionics components come from Rockwell-Collins. Communication between sensors, the computer, and data storage is by ethernet. A special variation, time triggered ethernet, is used to guarantee time of arrival of critical commands.

As has been pointed out in the Press, the computer onboard Orion is comparable to that of a contemporary cellphone. The state-of-the-art in spacecraft computers always lags the state-of-the art for commercial units, due to the harsh environment.

In the aircraft version, the "glass flight deck" uses a Rockwell Collins integrated display system with five 15.1-inch diagonal liquid crystal displays (LCDs), plus dual LCD head-up displays (HUD). The Common Data Network (CDN) from Rockwell Collins is a bi-directional copper and fiber optic network that utilizes ARINC 664 standards and protocols to manage the data flowing between the onboard systems. It is based on Ethernet technology and enabled for avionics systems. The CDN has higher data rates, expanded connectivity, and reductions in weight when contrasted with point to point topologies.

Each 787 aircraft contains has two Astronautics' Electronic Flight Bags (EFB's) onboard, that consist of a dual processor computer and display. The EFB hosts applications that provide pilot information via touch screen and bezel in a user-friendly digital format. The Astrionics EFB contains dual processors and dual hard drives to implement a hardware partition between the certified and uncertified applications.

SLS

The Space Launch System from Boeing has a triple redundant flight computer, with onboard touch screen displays for the crew.

The Soviet Experience

The Soviet spacecraft computer experience is less well documented and less accessible than the US effort. It began, as did the US's, in the 1960's, and was centered at the Research Institute of Research Machines. The family of computers was called the Argon.

The Soviets did not enjoy the established computer and integrated circuitry infrastructure that the US did. However, the laws of size, weight, power, and radiation tolerance are independent of political philosophy.

The Soviet Union placed the first artificial satellite of the Earth in orbit in 1957. Their rocketry and guidance followed the same path as the Americans, based on German World War-II experience. The development of Soviet space computers came about at the same time that MIT was developing the Apollo Guidance Computer. Earlier they used analog computers for control. The original Soyuz crewed craft was designed without an onboard computer. The Zond spacecraft had the first onboard digital computer, the Argon-11S, designed and built by the Moscow-based Scientific Research Institute of Electronic Machinery. In August 1969, the Zond-7 accomplished the first circumlunar mission, under control of the Argon. There were eventually 11 different Argon models, from 1964 to the mid 1970's.

Argon-11C was used on the Zond lunar probes. Triple Modular Redundancy was used. Argon-16 models were used extensively on Soyuz, Salyut, the Progress transports, MIR, and other missions. The Argon-16 is a 16-bit machine, with 6 kilobytes of random access memory, and 48 kilobytes of read-only memory. Sixteen interrupts are supported. It accommodates a wide variety of analog and digital Input/Output devices. It weighs 70 kilograms, and used 280 watts of power. It entered production in 1974, and is still being produced. Soyuz missions to the International Space Station use the Argon-16 computer.

The Argon-17 was radiation-hard. This was accomplished by a radiation detector, triggering a special circuit called the restarter (rebooter).

The Argon-11c was a fixed point, 14-bit (data) machine, with 17-bit instructions. There were 15 instructions. It could do an add in 30 milliseconds, a multiply in 160. It had 128 words of RAM, and 4096 words of ROM. It was built from integrated circuits. It weighed 34 kilograms, and consumed 75 watts.

The Argon-12S was delivered to the Mir Space Station, and took over attitude and orbit control of the 240 ton station. The first crewed Soviet spacecraft with a computer was the Soyuz T-2, in June 1980.

The Salyut series of onboard computers were designed and produced by the Scientific Research Institute of Micro-Instruments near Moscow. Eventually, the MIR space station used a series of computers, the Argon-16, the Salyut-4, and the Salyut-5.

When Soviet flight computers went into production, it was at the Kiev Radio Factory, a technology center in the Ukraine. It went on to produce over 300 flight computers. The facility was originally a railroad repair shop, but became a major computer design and development center. It originally produced missile guidance computers, which only have to operate for 10-15 minutes. It transitioned to design systems that would work for 100's and thousands of hours.

The Soviet Buran winged crewed vehicle had a sophisticated

system of dual 4-computer units, with voting logic. Unlike the Shuttle, the Buran could be flown and recovered in an unmanned mode. The flight computer, built in the Ukraine, was named the Sayiut. It featured 132k of ram, and 16k of ROM, using 36 bit words. There were two new programming languages developed for it, Prol-2 (onboard) and Dipol (ground). The Buran vehicle was never involved in a crewed flight, but was tragically destroyed when the roof of the hanger collapsed on it.

Unmanned spacecraft computers

The drivers for the development of programmable computers onboard the unmanned spacecraft were these:

- Autonomy – no need to rely on real-time or stored commands.

- Flexibility of operations – the ability to respond to unanticipated events.

Both of these features would reduce the cost of supporting missions post-launch. However, new support systems would be required with software tools such as an assembler, linker/loader, a debug facility, and a simulation capability. The initial space computers for unmanned satellites were purpose-built from the logic chips available at the time, much like the Apollo Guidance Computer. As technology advanced, monolithic microprocessors would be applied.

NASA Earth Orbiting Missions

Earth orbiting, non-crewed missions are the responsibility of NASA's Goddard Space Flight Center (GSFC) in Greenbelt, Maryland.

In 1971, the Interplanetary Monitoring Platform IMP-I mission was launched for magnetospheric research. (the author was in the control center at GSFC for this launch). It flew with a very early flight computer, the SDP-3. This was a 16-bit serial machine. It had both a user mode and a monitor (executive) mode. It implemented 54 instructions on single precision, 2's complement, fixed point data. Multiplication and division were done in software. It has 16 levels of interrupt. All instructions took the same time to execute, 78 microseconds. It had a 4k core memory unit, and consumed around 9 watts.

NASA's OBP

The On Board Processor (OBP) was the computer of the Orbiting Astronomical Observatory (OAO) Mission *Copernicus*, built by Grumman Aerospace in 1972.

It was developed by the Flight Data Storage Branch at GSFC, with Westinghouse fabricating the engineering model, Electronic Memories, Inc. providing the core modules, and the Branch doing the I/O and power converter. Flight units were ready by March of 1970. Some issues related to wiring were uncovered during testing, and this resulted in changes to the fabrication of the second unit. Acceptance testing for the OAO-C mission was accomplished by the end of August, 1970. After some rework, the OBP was integrated with the spacecraft in December, with Flight Acceptance test completion in March of 1971.

The OBP used some 1300 monolithic microcircuit packages. The engineering model used the Fairchild 9040 series diode-transistor logic chips. A test version was built on 138 wire-wrap circuit cards in a backframe. It needed 5 watts of power, with up to 30 additional watts for the core memory. It achieved a 6.25 microsecond add. It included an index register, and hardware multiply and divide. The CPU was approximately 8x6.5x4.5 inches, with each 4k of memory roughly the same. The design included interrupts, including a non-maskable one, and a real time clock.

It was considered experimental; i.e., not required for operations, but its use resulted in extension of the mission, and simplification of the real-time operations from the ground. The computer could be switched in to provide standard commands to the spacecraft, in place of uplinked commands. For safety, the relays that accomplished this would only remain in the "onboard" position for 37 seconds, and then switch back to the command receiver. The computer received spacecraft data intended for telemetry, and could interface with the stabilization & control system for attitude adjustment. The entire computer memory contents could be telemetered to the ground, or loaded from the ground.

The use of an onboard computer, even in an experimental fashion, was seen to be a powerful tool for operations. Although support of the computer itself was now added to the operations burden, the computer could monitor and command the spacecraft. Thus, operations could continue for the entire orbit, not just when the spacecraft was visible from ground stations. In addition, the OBP could make decisions onboard, and change operations to correspond to changing events.

After 4 ½ years on-orbit, the cpu suffered an ALU hardware problem, where Bit 6 of the adder was stuck. This was diagnosed from the ground down to the chip level. The problem was that the ALU was also used to increment the program counter. However, in an extraordinary effort, the onboard software was re-written to bypass the problem. (i.e., not to use bit 6). This was mostly done on a yellow lined pad, with a #2 pencil.

Early support systems for onboard computers evolved along with the flight systems. These were usually hosted on mainframe computers, with the source language on punched cards and magnetic tape. The programming language used was assembly (if not machine language), due to the very limited resources available. The development and verification environment used dedicated hardware simulators, which were very expensive, and custom interfaces. A major problem was verifying the fidelity of the simulator, particularly in terms of timing.

GSFC's support software system for the OBP consisted of a cross-assembler, a loader, a simulator, a library of software routines, and cpu diagnostics. The library included routines for arctangent, exponential, log, sine, cosine, and square root. The support software was written in Fortran, for portability. It was run on an SDS-920 computer as well as a Univac 1108 and GE 635. The 920 had a hardware interface to an engineering model of the OBP.

The code was produced on punched cards, and was compiled and output to magnetic tape as a memory image. The assembler could operated with base-10 or base-8 (octal) numbers. Scaled integers could be employed, with a global scaling factor.

The Advanced On-board Processor

The next model of a flight computer used on unmanned missions was the Advanced On-Board Processor (AOP), flown on Landsat-B and -C, IUE, and OSS-1 (which was a payload on STS-3). The AOP cycle time for the IUE mission was 1.3 microseconds. The computer had 12k of 18 bit words.

The goal of the follow-on Advanced Onboard Processor was to retain reliability, while increasing speed, and reducing power consumption. The goal was a 4 microsecond add time, 30 microsecond multiply, and 60 microsecond divide, with a total cpu and memory power usage of around 9 watts. The processor was built in TTL technology, and the memory was plated-wire magnetic.

IUE's AOP woes

The temperature specification for the onboard electronics was 38 deg C. The observed post-launch environment was 52 to 55 deg C. Thus the computer was operating in a realm that it had not been tested in.

On the mission, there were many strange computer crashes, leading to loss of attitude control. Nothing like this had ever happened in test, so it had to be thermally related. A memory dump analysis showed corrupted interrupt vectors. This was the first big clue as to what was happening. The personnel who had developed the software had been by this time reassigned to other projects.

There was no ground-based facility that could duplicate the OBC temperature seen in flight. However, software to provide protection against "hits" was completed 2 months post-launch, as a work-around. This seemed to solve the problem, and the mission continued to return valuable scientific data.

About 1.5 years after launch, there was a detailed study of the cause of the problem. Analysis revealed a fault condition after an ADD of two values, both close to zero, and both negative. An interrupt causes a context switch. When the adder circuit was hot, it took longer to settle. Since the adder was also used to generate the target address, (there was no separate program counter incrementer) sometimes, with previous negative numbers, bit 15 was not cleared. This resulted in a wrong interrupt jump target vector. A diagnostic patch to fix the problem was finally uploaded in January 1980.

NSSC-1

The NASA Standard Spacecraft Computer (NSSC-1) was developed at Goddard Space Flight Center as a general unit for a wide variety of spacecraft missions. It had an 18-bit word, and operated initially at 500 Khz. It was a 2's complement machine, with 50 general purpose instructions. A storage limit register provided a block of memory in which an operating system could reside, separate from the application programs. Memory modules initially were core, in units of 4096 words. Data I/O was independent of the cpu operation. It was possible to uplink new memory contents, and downlink (dump) memory contents to the ground. The computer ran a variety of application software, handling commanding and telemetry, monitoring of subsystems, and attitude control. There was a capability of sending a time-tagged command to the cpu, that would be sent to the spacecraft systems at the appropriate time.

The NSSC-1 was developed as a standard component for the

Multi-Mission Modular (MMS) Spacecraft at GSFC in 1974. The basic spacecraft was built of standardized components and modules, for cost reduction. The computer had 18 bit words using core or plated wire memory; up to 64 k. 18 bits was chosen because it gave more accuracy (x4) for data over a 16 bit machine. Floating point was not supported.

The STINT, or Standard Interface was the I/O component of the NSSC-1. It interfaced to the various spacecraft equipment.

The "Care and Feeding" of the onboard computers took on an increasingly important role. Ground support environments, usually hosted on mainframes, developed as the onboard systems became more sophisticated. An SDS-920 mainframe computer was used in the development and testing environment, and for support after launch. Programming was generally done in machine language. The development machine had a card reader, paper tape punch/reader, magnetic tape drive, a printer, and a console typewriter.

The NSSC-1 was used on the SMM, Hubble Space Telescope (HST), the Landsat missions, and UARS, among others. The hardware was developed by Westinghouse and GSFC. The machine had originally used DTL (diode-transistor logic), the lowest power parts available at the time on the Preferred Parts List. The NSSC-1 was later switched to 69 MSI (medium scale integration) chips. The NSSC-1 was also used on the ESA Sentinel-1 mission.

Hubble's Command and data handling (C&DH) module contains a Central Processing Module (CPM) that included the NSSC-1, four memory modules and a Standard Interface (STINT) unit, which served as the communications interface between the NSSC-1 and the Control Unit/Science Data Formatter, or CU/SDF. The flight software in the NSSC-1 computer monitors and controls the science instruments and the NICMOS Cooling System.

It was replaced on orbit during servicing mission 4, ST-125, in May 2009, because the "A" side had failed in 2008. It operated on the redundant "B" side until replaced.

Hubble's instruments and equipment also had their own microprocessors. The Multiple Access Transponder's used Hughes 1802's, as did the Wide field camera. This unit was replaced on-orbit twice.

NSSC-1 programming and support

A purpose-built NSSC-1 Flight Executive was developed and used on the Solar Maximum Mission (SMM) and subsequent flights. It time-sliced tasks at 25 ms. It included a stored command processor that handled both absolute time and relative time commands. It included a status buffer that could be telemetered back to the ground. It required a lot of memory, typically more than half of that available, leaving the rest for applications and spare.

The NSSC-1 had an Assembler/loader/simulator toolset hosted on Xerox XDS 930 (24-bit) mainframe. An associated simulator ran at 1/1000 of real time. The Xerox computer was interfaced to a breadboard OBP in a rack. Later, the Software Development and Validation Facility (SDVF) added a flight dynamics simulator hosted on a PDP-11/70 minicomputer.

A breadboard, or non-flight NSSC-1 and Stint was used in the facility. As the equipment aged, it became less reliable, yet was required to support ongoing missions. The author lead a team that suggested that the "simple" architecture of the NSSC-1 could be instantiated in an FPGA. It was eventually successful, but we learned the hard way, the same way the Penn State Group learned in implementing a FPGA-based Eniac, that a lot of the logic compensated for the idiosyncrasies of early logic. Once we figured out what was logically required, and got rid of what compensated for the use of the early NOR gates, we were good to go. In addition, a simpler FPGA-based STINT was developed, and was integrated into the C&DH simulation rack, for the Compton Gamma Ray Observatory (GRO) Mission.

DF-224

The Space Telescope used a more advanced flight computer called the DF-224 from Rockwell Autonetics for spacecraft control. The DF-224 was a 24-bit fixed point machine. It operated a 1.25 MHz, with 64 kilowords of memory. It was programmed in assembly language.

Rockwell Autonetics built the circa-1980's DF-224 space-rated computer. It included triple 24-bit cpu's, with multiple memory units of 8 k words each. The computer used plated wire memory, a non-volatile technology similar to core. There are also triple I/O units. The computer used 2's complement, fixed point arithmetic. Clock speed was 1.25 Mhz. It was rather heavy at 110 lbs.

The DF-224 was also considered for the guidance computer for the Athena Upper Stage, which would be launched from the Space Shuttle. That stage had a capability of 2 tons to Geosynchronous orbit, from the Shuttle in orbit. These were intended for Shuttle flights to Polar orbit, from the Vandenberg launch site in California. None of these were ever flown.

By 1992, two of the six memory units of the DF-224 had failed, and a minimum of three working units were needed for spacecraft operation. A co-processor was installed by the first servicing mission in 1993. It had dual redundant 80386/80387 processor/numeric processor pairs, each with 1MB of RAM and 256kB EEPROM, plus 384kB of non-alterable permanent ROM. It was designed at the Goddard Space Flight Center, and was added to the DF-224 to augment its capabilities. It connected by a shared memory interface. The in-orbit upgrade was possible because the DF-224 had accessible external electrical connectors. The upgrade by accomplished by astronaut extra-vehicular activity (EVA) operations.

The DF-224 was replaced altogether on the third servicing mission in 1999 by the Advanced Computer. This has three rad-hard Intel 486 processors running at 25MHz, each with 2MB of SRAM and 1MB of EEPROM. The replacement computer increased the performance by a factor of 20. This remains the main processor on the spacecraft, and is currently operating. With the retirement of the space shuttle, no further upgrades are currently possible.

ATS-6 DOC

The Application Technology Satellite–6, a TDRSS pathfinder, used the Digital Operations Controller (dual units) by Honeywell. For this mission, by management direction, there were to be no changes in the Flight Software post launch, and thus there was no provisions made for software debug, development, simulation, and test. That lasted for 1 sidereal day (23 h 56 m) after launch.

A software overflow condition discovered post-launch affected the sign of the roll control loop. It was a two's complement variable, and went from a small negative to a large positive number in one step. The problem was worked out as a machine language patch literally on back of an envelope, on the author's flight back from the West coast. The patch was uplinked and tested, and operated correctly.

Later in the mission, loss of a star sensor was countered by reprogramming to use simultaneous observations from two other orthogonal sensors. The flexibility of using a programmable flight system was validated.

NSSC-II

The architectural design of the NSSC-II was that of an IBM System/360 mainframe and supported all but four instructions of the *S/360* Standard Instruction Set. This unit was not used, but influenced the design of the Shuttle's 4-Pi systems.

Challenges for Spacecraft Computers

The space environment is hostile and non-forgiving. There is little or no gravity, so no convection cooling, leading to thermal problems. There is a high radiation environment. The system is power constrained. And, it is hard to fix stuff after launch.

There are differing environments by Mission type. For Near-Earth orbiters, there are the radiation problems of the Van Allen belts and South Atlantic Anomaly, and the issue of atmospheric drag. Missions tended to be Shuttle serviceable, as long as the Shuttle fleet was available. Synchronous or L2 (Lagrange Point) missions are not fixable, at the present time. There are ongoing efforts for servicing by robotic satellites, at GEO. If we go towards the sun it gets hot. That includes missions to Venus or Mercury. If we go away from the sun, it gets cold, and the amount of energy we can capture via solar arrays is limited. This includes missions to the Asteroids, Mars, Jupiter, Saturn, the outer planets, and their associated moons.

Planetary Probes include orbiters, rovers, and surface packages, Mercury and Venus landers (which tend to melt), Mars rovers and orbiters, Jovian and Saturnian moon probes, which have to deal with extreme radiation belts, and missions to the outer planets and beyond the solar system.

The functions of the computers on the spacecraft include attitude control and pointing, orbit control & maintenance, thermal control, energy management, and data management and communications (which may include antenna pointing).

ITAR

Systems that provide "satellite control software" are included under the *International Trafficking in Arms* (ITAR) regulation, as the software is defined as "munitions" subject to export control. The Department of State interprets and enforces ITAR regulations. It applies to items that might go to non-US citizens, even citizens of friendly nations or NATO Partners. Even items received from Allies may not necessarily be provided back to them. Software and embedded systems related to launch vehicles and satellites are given particular scrutiny. The ITAR regulations date from the period of the Cold War with the Soviet Union. Increased enforcement of ITAR regulations recently have resulted in American market share in satellite technology declining. A license is required to export controlled technology. This includes passing technical information to a foreign national within the United States. Penalties of up to $100 million have been imposed for violations of the ITAR Regulations, and imprisonment is also possible. Something as simple as carrying ITAR information on a laptop or storage medium outside the US is considered a violation. ITAR regulations are complex, and need to be understood when working in areas of possible application. ITAR regulations apply to the hardware, software, and Intellectual Property assets, as well as test data and documentation. It is a complex topic.

Radiation Effects

There are two radiation problem areas: cumulative dose, and single event. Operating above the Van Allen belts of particles trapped in Earth's magnetic flux lines, spacecraft are exposed to the full fury of the Universe. Earth's magnetic poles do not align with the rotational poles, so the Van Allen belts dip to around 200 kilometers in the South Atlantic, leaving a region called the South Atlantic Anomaly. The magnetic field lines are good at deflecting charged particles, but mostly useless against electromagnetic radiation and uncharged particles such as neutrons. One trip across the Van Allen belts can ruin a spacecraft's electronics. Some spacecraft turn off sensitive electronics every ninety minutes –

every pass through the low dipping belts in the South Atlantic.

The Earth and other planets are constantly immersed in the solar wind, a flow of hot plasma emitted by the Sun in all directions, a result of the two-million-degree heat of the Sun's outermost layer, the Corona. The solar wind usually reaches Earth with a velocity around 400 km/s, with a density around 5 ions/cm^3. During magnetic storms on the Sun, flows can be several times faster, and stronger. The Sun tends to have an eleven year cycle of maxima. A solar flare is a large explosion in the Sun's atmosphere that can release as much as 6×10^{25} joules in one event, equal to about one sixth of the Sun's total energy output every second. Solar flares are frequently coincident with sun spots. Solar flares, being releases of large amounts of energy, can trigger Coronal Mass Ejections, and accelerate lighter particles to near the speed of light.

The size of the Van Allen Belts shrink and expand in response to the Solar Wind. The wind is made up of particles, electrons up to 10 Million electron volts (MeV), and protons up to 100 Mev – all ionizing doses. One charged particle can knock thousands of electrons loose from the semiconductor lattice, causing noise, spikes, and current surges. Since memory elements are capacitors, they can be damaged or discharged, essentially changing state.

Vacuum tube based technology is essentially immune from radiation effects. The Russians designed (but did not complete) a Venus Rover mission using vacuum tube electronics.

Not just current electronics are vulnerable. The Great Auroral Exhibition of 1859 interacted with the then-extant telegraph lines acting as antennae, such that batteries were not needed for the telegraph apparatus to operate for hours at a time. Some telegraph systems were set on fire. The whole show is referred to as the Carrington Event, after British Amateur astronomer Richard Carrington.

Around other planets, the closer we get to the Sun, the bigger the impact of solar generated particles, and the less predictable they are. Auroras have been observed on Venus, in spite of the planet not having an observed magnetic field. The impact of the solar particles becomes less of a problem with the outer planets. Auroras have been observed on Mars, and the magnetic filed of Jupiter, Saturn, and some of the moons cause their "Van Allen belts" to trap large numbers of energetic particles, which cause more problems for spacecraft in transit. Both Jupiter and Saturn have magnetic field greater than Earth's. Not all planets have a magnetic field, so not all have charged particle belts.

Cumulative dose and single events

The more radiation that the equipment gets, in low doses for a long time, or in high doses for a shorter time, the greater the probability of damage.

These events are caused by high energy particles, usually protons, that disrupt and damage the semiconductor lattice. The effects can be upsets (bit changes) or latch-ups (bit stuck). The damage can "heal" itself, but its often permanent. Most of the problems are caused by energetic solar protons, although galactic cosmic rays are also an issue. Solar activity varies, but is now monitored by sentinel spacecraft, and periods of intensive solar radiation and particle flux can be predicted. Although the Sun is only 8 light minutes away from Earth, the energetic particles travel much slower than light, and we have several days warning. During periods of intense solar activity, Coronal Mass Ejection (CME) events can send massive streams of charged particles outward. These hit the Earth's magnetic field and create a bow wave. The Aurora Borealis or Northern Lights are one manifestation of incoming charged particles hitting the upper reaches of the ionosphere.

Cosmic rays, particles and electromagnetic radiation, are omni-directional, and come from extra-solar sources. Most of them, 85%, are protons, with gamma rays and x-rays thrown in the mix.

Energy levels range to 10^6 to 10^8 electron volts (eV). These are mostly filtered out by Earth's atmosphere. There is no such mechanism on the Moon, where they reach and interact with the surface. Solar flux energy's range to several Billion electron volts (Gev).

Other interesting problems plague advanced electronics off-planet. The Hughes (Boeing) HS 601 series of spacecraft suffered a series of failures in 1992-1995 due to relays. In zero gravity, tin "whiskers" grew within the units, causing them to short. The control processors on six spacecraft were effected, with three mission failures because both computers failed. This was highly noticeable, as the satellites were communication relays.

The effects of radiation on silicon circuits can be mitigated by redundancy, the use of specifically radiation hardened parts, Error Detection and Correction (EDAC) circuitry, and scrubbing techniques. Hardened chips are produced on special insulating substrates such as Sapphire. Bipolar technology chips can withstand radiation better than CMOS technology chips, at the cost of greatly increased power consumption. Shielding techniques are also applied. Even a small thickness of aluminum blocks many of the energetic particles. However, a problem occurs when a particle collides with the aluminum atoms, creating a cascade of lower energy particles that can also cause damage. In error detection and correction techniques, special encoding of the stored information provides a protection against flipped bits, at the cost of additional bits to store. Redundancy can also be applied at the device or box level, with the popular Triple Modular Redundancy (TMR) technique triplicating everything, and assuming the probability of a double failure is less than that of a single failure. Watchdog timers are used to reset systems unless they are themselves reset by the software. Of course, the watchdog timer circuitry is also susceptible to failure.

Thermal issues

Radiation is not the only problem. In space, things are either too hot or too cold. On the inner planets toward the Sun, things are too hot. On the planets outward of Earth, things are too cold. In space, there is no gravity, so there are no conduction currents. Cooling is by conduction and radiation only. This requires heat-generating electronics to have a conductive path to a radiator. That makes board design for chips, and chip packaging, more complex and expensive.

Mechanical issues

In zero gravity, everything floats, whether you want it to or not. Floating conductive particles, bits of solder or bonding wire, can short out circuitry. this is mitigated by conformal coatings, but the perimeter of the die is usually ground, and cannot be coated due to the manufacturing sequence.

The challenges of electronics in space are daunting, but much is now understood about the failure mechanisms, and techniques to address them.

IP-in-space

The use of Internet Protocol for space missions is a convenience, and piggy-backs on the large established infrastructure of terrestrial data traffic. However, there are problems. A variation of mobile IP is used, because the spacecraft might not always be in view of a ground station, and traffic through the Tracking and Data Relay Satellites involves a significant delay. A hand-off scheme between various "cell" sites must be used, and a delay-tolerant protocol.

The formalized Interplanetary Internet evolved from a study at JPL, lead by Internet pioneer Vint Cerf, and Adrian Hook, from the CCSDS group. The concepts evolved to address very long delay and variable delay in communications links. For example, the Earth to Mars delay varies depending on where each planet is located in its orbit around the Sun. For some periods, one planet is behind the Sun from the point of view of the other, and communications between them is impossible for days to weeks at a time.

The Interplanetary Internet implements a Bundle Protocol to address large and variable delays. Normal IP traffic assumes a seamless, end-to-end, available data path, without worrying about the physical mechanism. The Bundle protocol addresses the case of high probability of errors, and disconnections. This protocol was tested in communication with an Earth orbiting satellite in 2008

The CCSDS, Consultive Committee on Space Data Standards, has evolved a delay tolerant protocol for use in space.

The Interplanetary Internet uses a store-and-forward node in orbit around a planet (initially, Mars) that would burst-transmist data back to Earth during available communications windows. At certain times, when the geometry is right, the Mars bound traffic might encounter significant interference. Mars surface craft communicate to Orbiters, which relay the transmissions to Earth. This allows for a lower wattage transmitter on the surface vehicle. Mars does not (yet) have the full infrastructure that is currently in place around the Earth – a network of navigation, weather, and communications satellites.

The Cubesat Space Protocol is a network layer protocol, especially for Cubesats, released in 2010 It features a 32-bit header with both network layer and transport layer data. It is written in the c language, and works with linux and FreeRTOS. The protocol and its implementation is Open source. At the physical layer, the protocol supports CAN bus, I2C, RS-232, TCP/IP, and CCSDS space link protocol.

Hardware

Soon after their development, general purpose microprocessors were used in embedded roles. Eventually, most chip manufacturers produced special models of embedded chips, allowing the use of their standard software tools, but including lower power models with special capabilities for the embedded world. Many of these found application on spacecraft. The harsh environment of space imposed particular requirements on the chips, in the areas of vibration and shock tolerance for the launch phase, and temperature and radiation tolerance during the operational lifetime.

Microprocessor-based Space computers

<u>Flight Microprocessor Usage</u>

This is not by any means a comprehensive list.

Mission	cpu
Cassini	1750A
Clementine	1750A, 32-bit RISC
Cluster (ESA)	1750a
Coriolis	RAD6000
Cubesat	8051,PIC, Arduino, ARM
Deep Impact	RadLite 750
Deep Space-1	RAD6000
EO-1/Warp	Mongoose V
EOS Aqua	1750A (4) and 8051 (2)
EOS Aura	1750A (4) and 8051 (2)
ESO Terra	1750A (2)
EUVE	1750A
FAST	8085 (2)
FUSE	80386, 80387, 68000
Galileo	2900 (2) 1802 (19)
Galileo AACS	ATAC (bit slice)
Gravity Probe B	RAD6000
GRO	9900, LSI-11
HealthSat-II	80C186(2), 80C188

HESSI	RAD6000
HETE-2	M56001(DSP) (8), Transputer
HST	80386, 80486
Icesat, GLAS	Mongoose V
ICM	R3000
ISEE	8080
ISS	i80386, Pentium
Landsat-D	8x300, 8x305
Mangalyaan (Mars)	MA31750 (1750A)
MAP	Mongoose V, UTMC 69R000
Mars98	Rad 6000
Mars Climate Orbiter	RAD6000
Mars Observer	1759A
Mars Pathfinder	RAD6000
Mars Pathfinder Rover	80C85
Mars Polar lander	RAD6000
Mars Surveyor	1750A
Messenger-Mercury	RAD6000 (2)
Mighty Sat-II	TMS320C40 (DSP) (4)
MSTI-1,2	1750A
MSTI-3	1750A, R3000
New Horizons (Pluto)	Mongoose-V (4)
Pluto Express	32-bit RISC
PoSat-1	80c186, TMS320C25 and C30
Rosetta (ESA)	1750A, RTX2010.
Sampex	80386, 80387
SIRTF	RAD6000
SMEX	80386, 80387
SMEX-lite	RAD6000
SMM	6100A

Snap-1	StrongARM
Stardust	RAD6000
TDRSS, early	2901
Tiros, Block 5D	F-8

(see also
http://en.wikichip.org/wiki/List_of_microprocessors_used_in_spac
ecrafts)

A custom-built, TTL 4-bit processor was used on the Pioneer-10 Deep Space Mission, launched in 1972. There is some argument about whether this was an Intel 4004 cpu, or a custom build. The mission studied the asteroid belt, the solar wind, Jupiter, and the outer reaches of the solar system. The computer was used to hold, decode, and distribute commands transmitted from Earth. The mission lasted until 2003, when communications was lost due to distance, a mission duration of 30 years. As of March 2011, the spacecraft was some 102 Astronomical Units (AU= 93 million miles) from the Sun, where sunlight takes 14 hours to get to. The last successful reception of telemetry was on April 27, 2002; subsequent signals provided no usable data. The final signal from Pioneer *10* was received on January 23, 2003 when it was 12 billion kilometers (80 AU) from Earth. The backup spacecraft can be seen in the Smithsonian Air & Space Museum in Washington, D. C.

The 8-bit Intel 8080 chip found use about a variety of space missions, including NASA's OSS-1, 2, and 3 Shuttle-attached pallets, Hubble Space Telescope, International Sun-Earth Explorer, Seasat, and the French Meteosat program and OTS missions. The OSS attached pallets onboard the Space Shuttle were not exposed to the harsh environment of space for extended periods. The pallets accommodated multiple experiments and instruments. Seasat operated for 105 days in orbit, when it suffered a catastrophic electrical system failure.

The 8085 was an advanced version of the 8080. It had two new instructions to enable/disable three added interrupt pins (and the serial I/O pins). It also featured simplified hardware that required only a single +5V supply, and clock-generator and bus-controller circuits on the chip. It was binary compatible with the 8080, but required less supporting hardware, allowing simpler and less expensive microcomputer systems to be built. The 8085 found use in several space missions, including NASA's OSS series. It was also used on the 1997 JPL Mars Pathfinder Rover *Sojourner*. This Rover didn't stray far from its lander. The attitude control system on the WIRE spacecraft used an 80C85, as did the FAST and XTE missions.

The Intel 8051 8-bit microcontroller found application on NASA's environmental satellites Aqua and Aura. It is still being used, as a core, implemented in FPGA's. The UT69RH051 is a rad-hard 8051 from Aeroflex. It is rated for a total dose of 1 million rads, and is latch-up immune. It implements the 8051 instruction set, and has three 16-bit timers, on-chip ram, 32 I/O lines, supports 7 interrupts, and has an integral serial communications channel.

The RCA CMOS 1802 8-bit unit was used on JPL's Voyager, Viking, Ulysses, and Galileo space probes. Multiple units were definitely used on Galileo, but there is some question about its use on the other two spacecraft. Prior to Voyager, JPL was using simple flight computers, purpose-built, and not based on a microprocessor architecture. This Command Computer System (CCS) architecture was a custom 18-bit machine.

The Voyager spacecraft, previously called Mars-Jupiter-Saturn-77, were launched during a unique opportunity in 1977 to take them past the maximum number of outer planets. They went on to explore Jupiter, Saturn, Neptune, and Uranus before heading off to interstellar space. The Voyager's are now more than 13 light-hours beyond the Sun, and still returning data. In 2010, the returned data from Voyager-2 was garbled, leading to an investigation that showed the most likely cause was a flipped memory bit. Adjustments were made to the ground equipment to

67

accommodate this, and the spacecraft continues to return useful data some 33 years after launch. They are expected to return data through 2020. The main data computer was a custom 4-bit CMOS processor.

The 1802 was also quite popular with the builders of the OSCAR series of amateur spacecraft.

The Fairchild 8-bit found application onboard NASA's Payload Assist Module (PAM-D). This was, essentially, an upper stage for the space shuttle, allowing the shuttle to launch a payload into a geosynchronous or higher orbit. The PAM-D was a Delta-class solid-fuel module, providing the same performance to higher orbit as a Delta rocket launched from the ground. Since the operation of the PAM on orbit was only a matter of minutes, the usual issues of radiation resistance were not all that applicable. The associated support electronics and equipment for the launch was carried in the Shuttle bay, returned, and reused. It was also used on the Tiros weather satellites, and the similar military weather satellites, the Block5D.

A PIC-18 microcontroller was used on NASA's SuitSat-1 in February 2006. SuitSat-1 was an actual Russian spacesuit, beyond its useful life, instrumented, inflated, and cast off from the International Space Station. The controller board used a Microchip PIC18F8722 8-bit microcontroller, MCP9800 temperature sensor, and MCP6022 op amps.

The 12-bit Intersil 6100A was used on instruments on NASA's Solar Maximum Mission (SMM) spacecraft, launched in 1980. The chip used the DEC PDP-8 instruction set. It was low-power (about 50 milliwatts), and widely used in high-reliability applications. It was available in military-spec versions, and was dual sourced by Intersil. The radiation tolerance was 10^5 RADS.

Some members of Intel's 16-bit 8086 processors found application in spacecraft, using the low-power cmos version of the chip. The

HealthSat-II mission used two of the 80c186, and an 80c188. PoSat-1 used an 80c186. The TRMM spacecraft's ACE(attitude control electronics) used one.

A Texas Instruments 16-bit TI-9900 cpu was used on NASA/GSFC's GRO mission, the Compton Gamma Ray Observatory.

The DEC LSI-11 was used on the Shuttle attached OSS pallets, Hubble Space Telescope, and NASA's Gamma Ray Observatory (GRO) mission.

The MIL-STD-1750A lays out a formal definition of a 16-bit instruction set architecture. It does not specify an implementation. The standard allows for memory mapping up to 2^{20} 16-bit words. There are 16 general purpose registers. Some can be used as index registers, some as base registers. Any register can be used as the stack pointer. Both 16 and 32-bit integer arithmetic are supported, as well as 32- and 48-bit floating point. There is a new 1750B spec that allows for expansion.

There are many implementations of the 1750A architecture, including several that are built as radiation-hardened pieces. An example of one of these is the Dynex Semiconductor MA31750, used on NASA's GOES-N-O-P series of geostationary environmental satellites. The cpu uses 32 bit internal busses, and has a 32-bit ALU and 24 bit multiplier. It can address 64 k words of memory, but this can be expanded to 1 megaword with the MA31751 memory management unit.

The preferred language for the 1750A was Jovial, an Algol language variant; later, ADA and c were used as well. The 1750A is found in many aircraft and missile applications by the United States Armed Forces and their allies. A quick list of examples include the USAF F-16 and −18, the AH-64D helicopter, and the F-111. The architecture is also used by the Indian Space Research Organisation (ISRO), and the Chinese Aerospace industry. In

1996, the 1750A architecture was declared obsolete for future military projects.

The 1750A found applications in many space projects, including NASA's Earth Observation Satellites (EOS) Aqua, Terra, and Aura. It was used on ESA missions Cluster and Rosetta. JPL used seven of the processors on the Cassini Mission to Saturn, and more units on Mars Observer and Mars Global Surveyor. It was used on the Clementine spacecraft, a NASA-Naval Research Laboratory Program to study the Moon. The 1750A was deployed on the Johns Hopkins University Applied Physics Laboratory's MSX – Midcourse Space Experiment spacecraft, which used nine. The 1750A flew on EUVE, MSTI -1, -2, & -3, Landsat-7, NEAR, and is on the GOES-13, GOES-O, and GOES–P NOAA spacecraft. The SPOT-4 mission includes a F9450, a National Semiconductor implementation. GEC-Plessy also manufactures a radiation-hard RH1750A.

Seven of the BAE-manufactured 1750A's went to Saturn on the Cassini Mission. These are part of BAE's Advanced Spaceborne Computer module (ASCM). BAE claims 200 of these modules were in orbitas of 2011.

ESA funded the development of a space-rated 16-bit microprocessor in the early 1990's. Built by Dynex Semiconductor, the MA3750 was a multichip architecture built in CMOS/SOS technology, capable of 2 million instructions per second (mips) performance. It implements the MIL-STD-1553 architecture, and is now available in a single chip version.

The Intersil RTX2010 was a radiation-hardened 16-bit processor organized as a stack machine. The architecture supports direct execution of the Forth language. The RTX2010 was used in numerous spacecraft missions, including The Advanced Composition Explorer (ACE), the NEAR/Shoemaker mission, Timed, IMAGE (2000), instruments on AXAF, EOS, and EUV, MSX, XTE, Cassini, and MagSat.

It was also used on ESA's Rosetta comet mission, on the lander. It was a Harris RTX2010 16-bit microcontroller, programmed in Forth. It operated at 1.7152 MHz, with 16k prom/eeprom and 64k of static ram. There was also a 8000-series Actel FPGA used for I/O and as a watchdog.

JPL's Galileo spacecraft's Attitude and Articulation Control System (AACS) used an Applied Technologies Advanced Computer (ATAC), a 16-bit design. The machine had 2 kilobytes of ROM and 64 kilobytes of ram. It was programmed in assembly language and HAL/s, the higher level language developed for the Shuttle Program. It hosted a real-time operating system developed at Jet Propulsion Laboratory. It also had CMOS 1802's.

Signetics 8x300 parts formed the basis for electronics on Goddard's MMS (Multi-Mission Modular Spacecraft, including SMM, Landsat, and Hubble Space Telescope, and on the Shuttle-attached OSS pallets.

Eight of the Motorola 56001 were used on the HETE-2 mission. These were 24-bit Digital Signal Processors from Texas Instruments. The TMS320C40 flew on Mightysat-II (4 units).

The Space Station Mutiplexer/Demultiplexer uses a 80386SX with associated 80387SX floating point coprocessor with 16MHz clock. The SX model was 32 bits internally, but with a 16-bit external interface. Eight Megabytes of ram with EDAC was used. Serial digital and parallel channels, 1553 bus, optical channels, and a 300 megabyte mass storage device were employed.

The WIRE spacecraft used a 16 MHz 80386/80387 pair. WIRE had 1 megabyte of SRAM, 64 kilobytes of eeprom, and an 88 Megabytes of bulk memory. In addition, an 82380 DMA controller, and a 8251 serial controller were used. The FUSE spacecraft used an 80386/80387-16 MHz in its command and data handling, and also in its attitude control system, as did NASA's

Sampex, SMEX, TRACE, and SWAS missions. The XTE spacecraft used an 80386 in its data system.

The University of Surrey (UK) MicroSat series, including UoSat-12 used the 80386EX, an embedded version of the 80386. This circa 1994 chip had a static design, meaning the clock could be slowed or stopped without the microprocessor losing state.

The 80386EX model included the memory management features of the baseline 80386, and added an interrupt controller, a watchdog timer, sync/async serial I/O, DMA control, parallel I/O and dynamic memory refresh control. These devices were DOS-compatible in the sense that their I/O addresses, dma and interrupt assignments correspond with an IBM pc board-level architecture. The DMA controller was an enhanced superset of the 8237A DMA controller.

The 80386EX includes two dma channels, three channels of a 8254 timer/counter, dual 8259A interrupt controller functionality, a full-duplex synchronous serial I/O channel, two channels of 8250A asynchronous serial I/O, a watchdog timer, 24 lines of parallel I/O, and support for dram refresh.

The circa 1989 Intel 80486 is a follow-on to the 80386/387 chips, and most significantly incorporated the functions of the central processing unit (cpu) and floating point processors on one chip.

The 80486 was used on the Hubble Space Telescope. When the orbiting telescope's Control Unit/Science Data Formatter failed on the "A" side in 2008, NASA cautiously switched to the "B" side, which contained an 80486 processor in the Science Instrument Command & Data Handling Unit. The redundant components had not been powered on since 1990. The successful initiation of the "B" side of the redundant unit restored full data functionality to the spacecraft.

The Hubble Space Telescope's DF-224 on-board computer was

augmented in 1993 with a 16 MHz 80386/80387 coprocessor during the First Servicing Mission in 1993. Servicing Mission 3A replaced the DF-224 with a new computer based on the 80486. The DF-224 was a 1970's design, with triple redundant cpu's, only one of which was used at a given time. It was a 2's complement, fixed point machine, It included six memory units of 8k 24-bit words of plated wire memory.

The Department of Energy-Sandia Labs manufactured a radiation-hard version of the Pentium chip, under a no-cost license from Intel. Partners in the program included NASA, the Air Force Research Laboratory, and the National Reconnaissance Office.

The Coldfire processor from Freescale Semiconductor is a 32-bit embedded processor based on the Motorola 68000. They are available as licensable cores, and can be implemented in an FPGA.

General Dynamics makes a rad-hard version, the RH-5208. It is based on the Coldfire V2 architecture. It is an embedded microcontroller, with timers, memory, and I/O. The RH-5208 is used on NASA's Magnetospheric Multiscale Mission (MMS), launched in March of 2015. This involves four spacecraft, flying in formation.

The FUSE spacecraft used several Motorola 68020 processors in its instrument data subsystem, along with the associated 68882 floating point coprocessor.

The SeaStar satellite used three of the Motorola 68302 chips in a rad hard version. They were configured into a single master, multiple slave system. The 68302 was an embedded version of the 68000. It was termed an Integrated Multiprotocol Processor, or communications processor. It had a RISC core, and could support various communications protocols via downloaded firmware. It could also do Digital Signal Processing.

The Thor microprocessor, from Saab Ericsson Space, is a general-

73

purpose, single-chip 32-bit stack-oriented RISC-architecture. The microprocessor is intended for embedded computer systems with high performance requirements in real-time applications, combined with fast execution of programs written in Ada. The chip is implemented in a radiation-hard cmos process. Saab-Ericsson has supplied the processors for numerous European space missions and for the Ariane launch vehicles.

The Surrey Satellite Technology Nanosat Applications Platform (SNAP-1) was launched on June 28, 2000. The onboard computer (OBC) is based on Intel's StrongArm SA-1100 with 4 Mbytes of 32-bit wide EDAC protected SRAM. The error correction logic can correct 2 bits in every 8 using a modified Hamming code and the errors are "washed" from memory by software to prevent accumulation from multiple single-event upsets. There is 2 Mbytes of Flash memory containing a simple bootloader which loads the application software into SRAM.

The ERC32 is a radiation-tolerant 32-bit RISC processor for space applications. It was developed by Temic (now Atmel) for the European Space Agency. Two versions were manufactured, the ERC32 Chip Set (Part Names: TSC691, TSC692, TSC693), and the ERC32 Single Chip (Part Name: TSC695). These implementations follow SPARC V7 specifications. Cache and MMU functions are not included. Implementations went from a 3-chip set in the 1990's to a single chip version by the end of the decade. Support for the chipset version of the ERC32 has been discontinued. The LEON processor is the follow-on, and supports the SPARC V8 specification.

The VHDL models are distributed under the open source gnu lesser general public license. The architecture is supported by the languages Ada and c.

The Atmel AT697 is a SPARC Version 8 processor, implementing ESA's LEON2 fault-tolerant architecture. It includes the integer and floating point cpu's, caches, and a memory controller. It is

74

rated at 90 mips. It has a power consumption of under a watt, at 100 MHz. The earlier TSC695F product was based on the SPARC Version 7 architecture. These devices a rated for a total radiation dose of greater than 300 krad, and are latch-up immune to 70-90 MeV/mg/cm3.

The LEON project was started by the European Space Agency (ESA) in late 1997 to develop a high-performance processor to be used in European space projects. The objectives for the project were to provide an open, portable, and non-proprietary processor design, capable to meet future requirements for performance, software compatibility, and low system cost. To maintain correct operation in the presence of single event upsets (SEU's), extensive error detection and error handling functions were needed. The goals were to detect and tolerate one error in any register without software intervention, and to suppress effects from Single Event Transient (SET) errors in combinatorial logic.

The LEON family includes the first LEON1 VHSIC Hardware Description Language design that was used in the LEONExpress test chip developed in 0.25 μm technology to prove the fault-tolerance concept. The second LEON2 VHDL design was used in the processor device AT697 from Atmel and various system-on-chip devices. These two LEON implementations were developed by ESA. Gaisler Research, now Aeroflex Gaisler, developed the third LEON3 design and the fourth generation LEON, the LEON4 processor.

A LEON processor can be instantiated in programmable logic such as an FPGA or an ASIC. LEON processors are available as soft IP cores

All processors in the LEON series are based on the SPARC-V8 RISC architecture. LEON2(-FT) has a five-stage pipeline while later versions have a seven-stage pipeline. LEON2 and LEON2-FT are distributed as a system-on-chip design that can be modified using a graphical configuration tool.

The standard LEON2 includes an interrupt controller, debug support hardware, 24-bit timers, a UART, a 16-bit I/O port, and a memory controller.

The LEON3, LEON3FT, and LEON4 cores are typically used together with the GRLIB IP Library. While the LEON2 distributions contain one design that can be used on several target technologies, GRLIB contains several template designs, both for FPGA development boards and for ASIC targets that can be modified using a graphical configuration tool similar to the one in the LEON2 distribution. The LEON/GRLIB package contains a larger number of cores compared to the LEON2 distributions and also include a plug and play extension to the on-chip AMBA bus. IP cores available in GRLIB include a 32-bit SDRAM controller, and pci bridge with dma, a 10/100/100 ethernet MAC, 8/16/32-bit wide prom and sram controller, 16/32/64 bit wide DDR/DDR@ memory controllers, a USB 2.0 host, and a CAN bus, spi, I2C, ATA controllers, a timer, interrupt controller, and general purpose I/O port.

The term LEON2/LEON2-FT refers to the LEON2 system-on-chip design, which is the LEON2 processor core together with the standard set of peripherals available in the LEON2(-FT) distribution. Later processors in the LEON series are used in a wide range of designs and are therefore not as tightly coupled with a standard set of peripherals. LEON3 and LEON4 refers to only the processor core, while LEON/GRLIB is used to refer to the complete system-on-chip design.

The LEON2-FT processor is the single event upset (seu) tolerant version of the LEON2 processor. Flip-flops are protected by triple modular redundancy (TMR) and all internal and external memories are protected by error detection and correction (edac) or parity bits. Special license restrictions apply to this IP (distributed by ESA).

The LEON3 is a synthesizable VHDL model of a 32-bit processor

76

compliant with the SPARC V8 architecture. The model is highly configurable, and particularly suitable for system-on-a-chip (SOC) designs. The full source code is available under the GNU GPL license, allowing use for any purpose without licensing fee. LEON3 is also available under a proprietary license, allowing it to be used in proprietary applications.

There are several differences between the LEON2 processor models and the LEON3. LEON3 includes SMP support and a seven-stage pipeline, while LEON2 does not support SMP and has a five-stage pipeline.

The LEON3FT is a fault-tolerant version of the standard LEON3 SPARC V8 Processor. It has been designed for operation in the harsh space environment, and includes functionality to detect and correct single event upset (SEU) errors in all on-chip RAM memories. The LEON3FT processor support most of the functionality in the standard LEON3 processor, and adds Register file SEU error-correction of up to 4 errors per 32-bit word, Cache memory error-correction of up to 4 errors per tag or 32-bit word, Autonomous and software transparent error handling. There is no timing impact due to error detection or correction. It is a Static design, using a 2.5 volt supply. It supports the IEEE-754 floating point standard.

The LEON3FT core is distributed together with a special FT version of the GRLIP IP library. Only netlist distribution is possible.

A FPGA implementation called LEON3FT-RTAX is proposed for critical space applications. It is tolerant to a total dose of 3×10^5 rads, and can withstand SEU to 10^8 mev-cm2/mg.

In January 2010, the fourth version of the LEON processor was released. It has static branch prediction added to pipeline, optional level-2 cache, 64-bit or 128-bit path to AMBA AHB interface, and higher performance possible (claimed by manufacturer: 1.7

DMIPS/MHz as opposed to 1.4 DMIPS/MHz of LEON3).

The Real Time Operating systems (RTOS's) that support LEON are Rtlinux, PikeOS, eCos, RTEMS, Nucleus, ThreadX, VxWorks, LynxOS, POK, a free ARINC653 implementation released under the BSD license, and embedded linux.

An IDT R-3051 embedded controller was used on the Clementine Mission. Also called the "Deep Space Program Science Experiment." Launched on January 25, 1994, the objective of the mission was to test sensors and spacecraft components under extended exposure to the space environment and to make scientific observations of the Moon and the near-Earth asteroid 1620 Geographos. The Geographos observations were not made due to a malfunction in the spacecraft.

The RHC-3000 was based on a LSI Logic implementation of the MIPS-3000 cpu, implemented in rad-hard technology by Harris Corporation. It was hard to 1 megaRad, and had a low single event upset error rate. It ran a 20 MHz clock. It could use a floating point coprocessor. As a microcontroller, it had dual dma channels, dual counter/timers, a 50 Mbps serial I/O bus, and implemented error detection and correction on the main memory and cache.

The Analog Devices ADSP-21020, a 32 bit floating point digital signal processor, was manufactured in a rad-hard version by Lockheed-Martin in Manassas, Virginia.

The TX49 chip was introduced in 2001 as a 64-bit, radiation tolerant device. This was accomplished by the Hi-Reliability Components Corporation (HIREC) of Japan, under contract to the Japanese Space Development Agency. Toshiba developed the processor core IP in a 64 bit architecture. The host large scale gate array was developed by NEC. Kyoce developed the packaging.

Besides the MIPS instruction set, the chip had a 64-bit data bus, and a 36 bit physical address bus, implementing 64-bit virtual

78

addresses. There were 8k instruction and data caches with error detection and correction. The operating frequency was 25 MHz. The chip implemented floating point. It operated from 3.3 volts, in a temperature range of -40 to +85 degrees C. The packaging was a 302-pin ceramic flat pack.

The approach involved taking the RISC architecture, and implementing it in a inherently rad-tolerant FPGA.

One of the missions for the TX-49 was the Japanese Engineering Test Satellite-VII, or Kiku-8. It was launched in December of 2006 onboard an Ariane vehicle. It operated until January of 2017. It proved its worth during the earthquake and resulting Tsunami of 2011, providing images in real time.

The KomDIC-32 was a R-3000 compatible chip, manufactured by the Scientific Research Institute of System Development of the Russian Academy of Sciences. It had an integrated R3010 floating point unit. At least 14 models were produced, from before 2005 to 2016. There was a 64-bit MIPS version, the KOMDIV-64 (2008-2016), dual-issue, superscalar. A much later version was produced in silicon-on-insulator technology in a 675 pin chip, and had RapidIO, ethernet, and pci.

Mongoose

The Mongoose-V is a 32-bit microprocessor for space applications It was a radiation-hardened 10–15 MHz version of the MIPS 3000 architecture. The Mongoose was developed by Synova, Inc., with support from the NASA's Goddard Space Flight Center, Code 735. It was rated at 7.5 MIPS, and included floating point support. It was fabricated in Silicon-on-Insulator technology.

The Mongoose V processor is a space-rated derivative of the LR-3000 processor of MIPS heritage. It included a 4-kbyte instruction cache and a 2-kbyte data cache, as well as floating-point capability. However, the omission of the memory management unit forced the use of a flat memory model and precluded one of the more powerful features of advanced operating systems.

Features of the Mongoose included the MIPS R3000 Instruction Set, the R3010 Floating-point Unit, On-Chip 2KB Data Cache and 4KB Instruction Cache, and speed grades of 10MHz and 15MHz. On-chip Peripherals included Error Detection & Correction, dual Memory Protection, timers, a Dual UART, Expansion Interrupts, a waitstate Generator, and a DRAM Controller.

Radiation Hardness was specified to be LET > 80 MeV- cm2 / mg, and the unit is Latch-up immune. It was built with a silicon-on-insulator technology.
The Mongoose-V processor first flew on NASA's EO-1 spacecraft launched in November, 2000, where it functioned as the main flight computer. A second Mongoose-V controlled the satellite's solid-state data recorder.

Other spacecraft that use the Mongoose-V include:

- NASAs Microwave Anisotropy Probe (MAP), launched in June 2001, carried a Mongoose-V flight computer similar to that on EO-1. The spacecraft measures differences in the temperature of the Big Bang's remnant radiant heat, called the Cosmic Microwave background radiation across the full sky

- NASA's X-ray Timing Explorer (XTE) mission, as the instrument telemetry controller.

- NASAs Microwave Anisotropy Probe (MAP), launched in June 2001, carried a Mongoose-V flight computer similar to that on EO-1. The spacecraft measures differences in the temperature of the Big Bang's remnant radiant heat, called the Cosmic Microwave background radiation across the full sky.

- NASA's Space Technology 5 series of microsatellites

- NASA's IceSat Mission, launched into polar orbit in January 2003

- TIMED, or Thermosphere Ionosphere Mesosphere Energetic and Dynamics mission, was launched to study the dynamic of the Mesosphere and Lower Thermosphere portions of the Earth's atmosphere. The Mongoose served as the Control and Data Handling (C&DH) computer. It was launched in December 2001.

- EO-1, the Earth Observing Mission, used a Mongoose. It was launched in 2000, and is currently in extended mission mode.

- New Horizons Pluto Mission uses four 12 MHz Mongoose processors in its Command & Data Handling subsystem. The spacecraft was launched in 2006, and is past Pluto It went past Jupiter in 2007, and Saturn in 2008. It arrived at Pluto in 2015, and is continuing into the Kuiper Belt. Probably due to excessive radiation in the vicinity of Jupiter's magnetotail, on March 19, 2007 the Command and Data Handling computer experienced an uncorrectable memory error and rebooted itself, causing the spacecraft to go into safe mode. The craft fully recovered within two days, and is now in extended mission mode, heading outside the solar system.

The RH3000 family from Harris features the MIPS architecture, and is radiation hard to a level greater than 1 MegaRad. The two-chip set includes the CPU, floating point unit, timers, bus interfaces, DMA controller, power management, and serial channels. The core logic is used under license from LSI Logic.

At the box level, Harris can put up to 21 processors with 2 gigabytes of RAM in a product called CrossStar. The chipset is also placed on a single 6U VME board called the Standard Spacecraft Processor Module. The Advanced Spacecraft Processor Module includes two processor sets and memory.

The product is supported by the VxWorks operating system and Tornado, and can be programmed in the c language or Ada.

RH32

The RH-32 was a radiation-hard 32-bit processor chipset developed by the USAF Rome Laboratories for the Ballistic Missile Defense Agency, and produced by Honeywell (later, TRW) for Aerospace applications. It achieves a throughput of 20 MIPS. It was a three-chip set, consisting of Central Processing Unit, Floating Point Unit, and Cache Memory

The Honeywell and TRW RH32 were developed from a MIPS R3000 model, under sponsorship of the USAF Phillips Lab at Kirkland Air Force Base in New Mexico. It features 16 kilobytes of data cache and 16 kilobytes of instruction cache. It includes four serial I/O channels, four timers, a built-in 1553 bus, 40 program-mable I/O lines, and DMA capability. At a module level, the Sun M-bus is supported. The module is available in 100 K rad to one mega-rad hardness with no single-event latch-up. It incorporates IEEE-754 floating-point capability, and memory management features. The RH32 processor is an integral part of the Advanced Spaceborne Computer Module (ASCM). The RH32 is supported by the VxWorks operating system and the gnu-c compiler.

Non-radiation hardened Motorola 603e's were used on NASA's Callipso (Cloud Aerosol Lidar and Infrared Pathfinder Satellite Observations) spacecraft. Four of the COTS units were used. for non-mission critical circuitry, many of the mitigation techniques were employed, costing significantly less than a completely radiation-hard part.

A new player in the game is the RISC-V architecture. Pronounced RISK-Five, RISC-V is the latest implementation of the MIPS architecture, in an open source configuration. The project kicked off in 2010 at the University of California, Berkeley.

The architecture was announced in 2018. Multiple companies now offer RISC-V chips, and the architecture is compatible with several FPGA's. The RISC-V architecture is making inroads on the popular ARM embedded architecture. What we're talking about here is an instruction set architecture, and a cpu specification. It is open source, meaning you can download it and build your own chips with it. If you implement it in a rad hard chip or FPGA, you have a rad-hard processor. Beyond the basic MIPS architecture, you can expand it to include the industry MMX instructions for video processing, digital signal processing, and the SIMD feature, for parallel processing. Using open source software, you can implement a real-time operating system, and use NASA/GSFC open source Core Flight Software for the applications. You get the architecture for free, and the software as well. The only expensive part is the rad-hard fpga, and its programmer and toolkit. Some commercial RISC-V chips are emerging. The RISC-V can be made Arduino compatible, making use of this microcontroller's vast ecosystem of hardware and software.

The Arduino HiFive-1 is a RISC-V board from Arduino.org. It was announced in 2017. It implements the RISC-V 32-bit architecture in a SiFive E310 SoC. It runs at 320 MHz, with an embedded wifi and bluetooth capability. So, what we have is a 32-bit Arduino architecture presented to the user, while inside, all that runs on a RISC-V MIPS architecture. The Arduino user does not need to be aware of the guts – not an Arm but a MIPS architecture. In building and running code, it is an "Arduino," albeit on steroids.

The TX49 chip was introduced in 2001 as a 64-bit, radiation tolerant device. This was accomplished by the Hi-Reliability Components Corporation (HIREC) of Japan, under contract to the

83

Japanese Space Development Agency. Toshiba developed the processor core IP in a 64 bit architecture. The host large scale gate array was developed by NEC. Kyoce developed the packaging.

Besides the MIPS instruction set, the chip had a 64-bit data bus, and a 36 bit physical address bus, implementing 64-bit virtual addresses. There were 8k instruction and data caches with error detection and correction. The operating frequency was 25 MHz. The chip implemented floating point. It operated from 3.3 volts, in a temperature range of -40 to +85 degrees C. The packaging was a 302-pin ceramic flat pack.

The approach involved taking the RISC architecture, and implementing it in a inherently rad-hard FPGA. That's a game changer.

One of the missions for the TX-49 was the Japanese Engineering Test Satellite-VII, or Kiku-8. It was launched in December of 2006 onboard an Ariane vehicle. It operated until January of 2017. It proved its worth during the earthquake and resulting Tsunami of 2011, providing images in real time.

KOMDIV-32

This was a R-3000 compatible chip, manufactured by the Scientific Research Institute of System Development of the Russian Academy of Sciences. It had an integrated R3010 floating point unit. At least 14 models were produced, from before 2005 to 2016. There was a 64-bit MIPS version, the KOMDIV-64 (2008-2016), dual-issue, superscalar. A much later version was produced in silicon-on-insulator technology in a 675 pin chip, and had RapidIO, ethernet, and pci.

RHPPC

The U. S. Air Force funded the PowerPC 603e processor's use in space. This was a development effort at the Southwest Research Institute, and the SC603e was produced by Honeywell and Thompson. It was a 100-mips class chip, hard to a total dose of 60 kilorad. The PowerPC 603 series was also used in the General Dynamics Integrated Spacecraft Computer (ISC).

The RHPPC is a radiation hardened cpu based on the PowerPC 603 technology licensed from and manufactured by Honeywell Space Systems. It was a project of the United States Air Force Research Laboratory at Kirtland Air Force Base. The RHPPC is equivalent to the commercial PowerPC 603e processor with the minor exceptions of the phase-locked loop (PLL) and the processor version register (PVR). The RHPPC processor is compatible with the commercial PowerPC architecture, the programmers interface, and is supported by common PowerPC software tools and embedded operating systems, such as VxWorks.

The RHPPC processor provides 190 mips with its core clock at 100 MHz. The RHPPC runs with a 25, 33.3, 40, or 50 MHz clock (SYSCLK) which is generated based on the PCI clock. The 60x bus clock is de-skewed on-chip by a PLL and can also be multiplied.

The RHPPC processor is a superscalar machine with five execution units: system register unit, integer unit, load/store unit, floating point unit, and branch processing unit. The dispatch unit can issue two instructions per cycle. The floating point unit has a three level deep pipeline. Out-of-order execution is supported through the use of shadow registers. The completion unit can complete two instructions per cycle in order by copying results from the shadow registers to the real registers. Independently, the branch processing unit can complete a branch each cycle. Thus, in theory, the RHPPC processor can complete three instructions per cycle. The RHPPC processor has a 16 kilobyte instruction cache and a 16 kB data cache that are 4-way set associative, and supports the write-through or copy-back protocol. A cache line is fixed at eight words.

The RHPPC processor is fabricated by Honeywell. It is packaged in a hermetic, 21 x 21 mm grid array package with 255 (16 x 16) leads.

Several versions of the PowerPC are also available in rad-hard versions. The Neutron Star Interior Composition Explorer (NICER) mission, launched in 2017, uses a BRE PowerPC 440. According to the company website, "The device is built on Honeywell's rad-hard HX5000 150nm SOI process and is based on a PowerPC 440 core licensed from IBM. The BRE440 processor design includes on-chip floating point unit, memory controllers, L2 cache, two Ethernet ports, two serial ports, four DMA channels, DDR DRAM controller with EDAC, and PCI interface. Speed is configurable up to 133MHz and is 2 MIPS/MHz." The Global Ecosystem Dynamics Investigation (GEDI) mission also uses this flight cpu.

RAD-6000

The RAD-6000 is a 32-bit version of the IBM PowerRISC, that was the processor in the Deep Blue computer that triumphed over human chess champion Garry Kasparov. IBM's RS/6000 series of circa-1990's Unix servers used the proprietary POWER architecture, later switching to the PowerPC. POWER stood for Performance Optimization With Enhanced RISC. This represented an Instruction Set Architecture, which came out of a superscalar design of the late 1970's.

The RAD6000 radiation-hard single board computer based on the IBM risc chip, was manufactured by IBM Federal Systems. Later sold to Loral and by way of acquisition, it ended up with Lockheed Martin and is currently as a part of BAE Systems.

The RAD 5545 was a product of BAE Systems, and is a multicore system-on-a-chip, using the POWER architecture. It is rad-hard by design, and is specified as good to a mega-rad, total ionizing dose.

The SRAMS are subseptable to single event upsets at the level of 2^{-9} per day. The chips uses less than 18 watts. It includes 4 RapidIO channels, a memory interface, and four RAD5500 processors, cache, I2C and SPI I/O interfaces, and a 16-port SpaceWire router. It is implemented in a 1,752 pin package. It dissipates a little under 18 watts.

The radiation-hardening of the original RSC 1.1 million-transistor chip to make the RAD6000's CPU was done by IBM Federal Systems Division working with the Air Force Research Laboratory. The memory bus is 72 bits wide, to accommodate 64 data bits plus 8 bits of error correcting code. The chip includes a floating point unit. It is latchup immune, and rated at a total dose exceeding 1 megarad. The architecture was implemented in a standard cpu, or a ASIC or FPGA version. Computer boards were available in double or triple-redundant configurations. Each processor drew less than 12 watts of power at 20 MHz. I/O interfaces included 1553 bus, a serial (RS-422 or -232) UART, and discretes in and out.

As of June 2008, there were some 200 RAD6000 processors in space on a variety of NASA, DoD, and commercial spacecraft, including:

The Spirit and Opportunity Mars Rovers, the Mars Pathfinder Lander, the Deep Space-1 probe, Mars Polar Lander and Mars Climate Orbiter, Mars Odyssey Orbiter, the Spitzer Infrared Telescope, the Messenger probe to Mercury, the Stereo spacecraft, MIDEX (Explorer 78), Genesis and Stardust sample return missions, the Phoenix Mars Polar lander, the Dawn asteroid belt mission, the Solar Dynamics Observatory spacecraft, launched Feb 11, 2010 flying both RAD6000 and RAD-750, Coriolis, Gravity Probe B, HESSI, SMEX-lite, SWIFT, Triana, and SIRTF.

The computer had a maximum clock rate of 33 Mhz and a processing speed of about 35 mips. In addition to the CPU itself, the RAD6000 has 128 megabytes of error-correcting RAM

VxWorks is supported on the RAD6000. The flight boards have selectable clock rates of 2.5, 5, 10, or 20 MHz.

The RAD750 is a radiation hardened single board computer based on a licensed version of the IBM PowerPC 750 architectture. It is specifically intended for use in high radiation environments in space. The RAD750 was released for purchase in 2001 and the first units were launched into space in 2005. Software developed for the RAD750 is upwardly compatible with the RAD6000, which is a Power PC architecture.

The cpu has 10.4 million transistors compared with the RAD6000's 1.1 million. It is manufactured using either 250 or 150 nm photolithography and has a die area of 130 mm² It has a core clock of 110 to 200 MHz and can process at 266 MIPS or more. The CPU includes a 32K instruction, and 32k data cache. An extended Level 2 cache to improve performance is optional . Its packaging and logic functions are completely compatible with the standard PowerPC 750.

The CPU itself can withstand 2,000 to 10,000 gray (1 gray = 100 rad) and temperature ranges between −55 and 125C. It requires 5 watts. The standard RAD750 single-board system (CompactPCI board form factor) can withstand 1,000 gray, temperature ranges between −55 °C and 70 °C and requires 10 watts of power. The clock ranges from 110 to 200 Mhz.

SpaceDev's HPX-21 computer uses dual redundant MPC750 Power pc's, and has been flown on CHIPSat. It is built around the compactPCI bus system, and hosts the VxWorks operating system.

The RAD750 can be purchased as a chip, or on a 3U CompactPCI board. This hosts a bridge ASIC to the memory and the PCI bus. A startup ROM for VxWorks is included. The PowerPC bridge provides a UART, and a JTAG connection. It also implements memory scrubbing.

88

There are many spacecraft in operation using RAD750 computers, including:

- The Deep Impact mission to a comet, launched in January 2005, was the first to use the RAD750 computer.

- The MARS Reconnaissance Orbiter (MRO) spacecraft has a 133 MHz RAD750 in the Command and Data Handling subsystem..

- The XSS-11 small experimental satellite.

- The Fermi Gamma-ray Space Telescope, formally GLAST, launched June 11, 2008

- Two RAD750 computers are in the World View-1 satellite which provide high-resolution imaging of Earth. It is a part of the National Geospatial Intelligence Agency's NextView program. It was launched in 2007.

- The Keppler Space Telescope, launched in March 2009.

- The Lunar Reconnaissance Orbiter (LRO) launched in June 2009.

- The Wide Field Survey Explorer (WISE) launched in 2009.

- The Titan-Saturn System Mission (TSSM) spacecraft will have a 132 MHz RAD750 on board, with a scheduled launch in 2020.

- The Solar Dynamics Observatory (SDO) spacecraft, l launched Feb 11, 2010, includes both RAD6000 and RAD-750 processors.

- The Global Precipitation Measurement spacecraft, GPM, launched in 2014.

- The Atlas instrument on Icesat-2, launched in 2018, has a RAD750 and three Leon3FT units.

- The follow-on to the Hubble, the James Webb Space Telescope, uses the RAD750 in the Instrument module.

- The geostationary environmental satellite, GOES-R, built by Lockheed Martin for NASA/NOAA,

- The Dscovr (Deep Space Climate observatory) spacecraft, launched in 2015.

- The Juno Mission, to Jupiter.

It was estimated that, by 2010, there were over 200 RAD750s in space

Other Earth Orbiters

The HETE-2 spacecraft, a follow-on to replace the lost HETE-1 spacecraft, was launched on Oct. 9, 2000. The spacecraft computer system consists of four identical processor boards: each board contains one T805 Transputer, two Motorola 56001 digital signal processors (DSP's), and 20 Megabytes of RAM. The processors are assigned to the spacecraft and science needs. The Transputer links allow for quick and efficient communications between processors. The digital signal processors (DSP's) serve as the interface to the instruments.

The embedded instrument controller for the Coronal Diagnostic Spectrometer was prototyped at GSFC using a T222 Transputer. This instrument was a part of the SOHO/ISTP (Solar Helispheric Observatory / International Solar-Terrestrial Physics) Program, jointly funded by the European Space Agency. This work was done as part of the Laboratory for Astronomy and Solar Physics.

The flight unit, using a T800 processor, was delivered for a Summer 1995 launch. The Transputer serves as an embedded controller, orchestrating the operation of the 1024 x 1024 element CCD (charge-coupled device) sensing element. The data system on the SOHO spacecraft also uses Transputers.

Early JPL missions

For the Mariner Venus spacecraft of 1962, a non-programmable Central Computer and Sequencer (CCS) was used. Around the same time, a UCLA professor, Algirdas Avizienis, was developing his concept of STAR, the self-testing and repair computer. Redundancy was used, but the spare units were unpowered unitl they were needed. This increased reliability through dormancy. It was a 32-bit, fixed point machine, with 16k or ROM. This was similar to what was used on the Apollo Guidance computer, with magnetic core. The key was the Testing and Repair Processor, TARP. This unit had three active modules, and two spare. Support software was developed on a Univac-1108 mainframe. In the end, STAR never flew. It was a model, however, for some of the architecture for the NASA Standard Spacecraft Computer-1.

It was considered for the Thermoelectric Outer Planet Spacecraft (TOPS)

Maxwell SCS-750 Spacecraft Supercomputer

This unit uses commercial grade, silicon-on-insulator PowerPC 750FX cpu's to achieve high performance and radiation tolerance. The silicon-on insulator-technology gives a design that is immune to latchup and highly tolerant to single event upsets. . The computer is a triple modular redundant configuration. Their is 512kbytes of level 2 cache, with error correcting codes. There is 256 megabytes of SDRAM and 64 Mbytes of flash, with advanced error detection and correction circuitry. The computer includes 1553 interfaces, a cPCI bus, 2 serial and 2 parallel ports, and EDAC. Power ranges from 5 to 20 watts for the box, and is software-selectable. Power consumption is kept low compared to a conventional rad-hard design. Part of the circuitry is implemented in an Actel SEU-immune FPGA.

The High Performance Spacecraft Computing (HPSC) Project was kicked off in 2012 by NASA. It is supported by engineers from NASA Centers JSC, GSFC, KSC, and JPL. It is part of the Game Changing Development Program of the NASA Office of the Chief Technologist, Space Technology Program. The goal is to advance the state-of-the-art in Rad Hard units that will use the commercial features of multicore, higher clock rates, Digital Signal processing, and other features that we find on our phones, but not on our spacecraft.

The Boeing Company was chosen to implement the concept. This is ongoing through 2020. The Chiplet, as the processor is called, will run linux and real-time operating systems. Each chiplet has a dual quad-core architecture. It is expected that the unit will have one hundred times the performance of the RAD-750. This program is managed by JPL.

Reference: http://gcd.larc.nasa.gov/projects/high-performance-spaceflight-computing

Spacecraft-on-a-chip

In the limit, the spacecraft can be implemented on a single chip. The computer is the vehicle. Cubesats, which are small payloads that colleges or even individuals can deploy into space, are run with what is essentially cell phone technology. Think about it – your cellphone probably has a 3-axis accelerometer, a magnetometer, a GPS unit, a gyroscope, a camera, and a high end embedded processor. A phonesat went into orbit in 2013, using the Google Nexus One. It was integrated into a 3-unit (3U) Cubesat called STRaND-1 from Surrey Space Technology.

But we can get smaller than this. As the electronics gets

implemented on denser and denser structures, we can get the processor, memory, input-output, and MEMS sensors on one piece of silicon. They can also economically be be produced in quantity, and deployed in swarms. They would be disposable. This concept is under development at numerous locations.

An architecture for ChipSat's was developed by Surrey Space Technology, UK. This was implemented in an Xilinx FPGA core, and included support for CCSDS telemetry and command protocols. It scopes the onboard data handling function for a small spacecraft. It includes a 32-bit risc core, and image processing hardware for video The project targeted the LEON microprocessor core and a CAN bus core, with an EDAC unit included. There is a CORDIC coprocessor. The CCSDS function is implemented in software. The US Air Force has an ongoing interest in satellites-on-a-chip.

In 2016, NASA and the U.S. Air Force approached industry about a new, rad-hard ARM processor. This was termed the High Performance Spacecraft Computing chip. It is based on the ARM A53 architecture, manufactured in a rad-hard-by design cell libraries. The A53 includes the NEON single instruction multiple data (simd) processor. In various architectural trade study's, a rad-hard, general purpose multicore was the chosen architecture. This will be produced by a rad-hard-by design methodology, and will incorporate Serial RapidIO.

A Rad-Hard ARM Cortex-M0 is a microcontroller from Protec GmbH, a company with 30 years experience in rad-hard electronics, and a portfolio of processor and support parts. It provides a Cortex M0 cpu, operating at 50 Mhz, and using 3.3 volts. Memory includes 16k each of data and program memory. Error detection and correction is included. It can interface with up to 36 megabytes of external memory. It includes GPIO pins, that can also be used as interrupts. There are 32 general purpose counter-timers, and dual UARTS. There are dual SPI interfaces.

All internal registers are triple-modular redundant. It is hardened to a TID of 300 krad, and is latch-up immune to 100 MeV-cm^2/mg. It comes in a 1.3 x 1.3 inch, 188 pin package. Not a computational powerhouse, it is a capable controller.

The SAMRH71 is a rad-hard-by-design microcontroller chip from Microchip, based on their commercial grade SAMv71. It is a 32-bit Cortex M7. The rad hard version includes Spacewire and MIL-STD-1553 I/O. The radiation performance is a LET of 62 MeV/CM^2 with an SEU greater than 20 and a TID of 100 Krad. This microcontroller operates up to 100 MHz. Besides the rad-hard part, a less-expensive radiation-tolerant part is also available. The part includes CAN and Ethernet interfaces.

Vorago Technologies is marketing the VA10820, an ARM Cortex-M0 that is hard to 300krad. It operates at 50 MHz, and includes JTAG. It has 32k of data memory, and 128k of program memory. Memory EDAC and scrub are included. It is a microcontroller, with 54 GPIO's, dual i2c, dual UART, and triple SPI's. It has a remarkable 24 32-bit counter-timers, and includes PWM's, and a watchdog timer. All internal registers are TMR.

Software

Software is the ideal component to use in space. It doesn't need radiation shielding, it doesn't weight anything. What's the worst that could happen?

Early flight software was written in assembly, using a small mainframe system to develop and test code. Each architecture was unique. The software tools had to be developed each time. As hardware began to be standardized, and particularly when it began to be a rad-hard version of standard commercial products, there was a surge in productivity. Flight software was developed in ADA, c, and HAL/S. Commercial toolsets were used. It was getting easier, at the time when more and more work was being done in software. The software became more and more complex.

Flight Software is a special case of embedded real-time software. There is generally no direct human interface such as monitor and keyboard/mouse. All interactions are through uplink and downlink, with delays based on the distances involved. The interfaces of the computer are with numerous specific spacecraft components; sensors such as star trackers, Earth sensors, sun sensors, temperature sensors, etc. The interface with actuators includes various mechanisms, thrusters, reaction control wheels, and actuators and positioners for science instruments. Flight software executes on radiation-hardened processors and microcontrollers that are slower than the commercial state of the art, and memory is more limited. This is a source of incidental complexity to the design. Most onboard computers are real-time systems, with synchronous and asynchronous timing requirements, and hard deadlines. It's an environment where being late is being wrong.

HAL/S was the high order assembly language for the Shuttle. It was produced in the early 1970's by Intermetrics for NASA. It was based on the PL/1 language. PL/I is a procedural computer programming language designed for scientific, engineering, and systems programming applications. It has been used by various academic, commercial and industrial organizations since it was introduced in the 1960s.

HAL/s was designed to be reliable, efficient, and machine-independent. It was optimized for aerospace-type applications, and included matrix operations. Certain attributes that could lead to problems were deliberately left out of the language, such as dynamic memory allocation. It includes support for floating point, vectors, matrices, strings, and boolean variables.

NASA didn't have a policy on use of Open Source software for quite a while. There were misunderstandings and suspicion about Open Source by government agencies and commercial firms. In the early 2000's, that began to change, as organizations began to realize the advantages of open source and collaborative development and testing environments. The current NASA open Source License, version 1.3, can be found here:

http://opensource.org/licenses/NASA-1.3

New start-ups such as SpaceX have embraced Open Source from the very beginning. Two of the Open Source projects for flight code are discussed below – the FlightLinux Project, and NASA/GSFC's Core Flight Executive/Core Flight Software.

Flight Computer housekeeping tasks

Besides attitude determination and control, the onboard embedded systems has a variety of housekeeping tasks to attend to.

Generally, there is a dedicated unit, sometimes referred to as the Command & Data Handler (C&DH) with interfaces with the spacecraft transmitters and receivers, the onboard data system, and the flight computer. The C&DH, itself a computer, is in charge of uplinked data (generally, commands), onboard data storage, and data transmission. The C&DH can send received commands directly to various spacecraft components, or can hold them for later dissemination at a specified time. The C&DH has a direct connection with the science instrument(s) for that data stream. If the science instrument package has many units, there may be a separate science C&DH (SC&DH) that consolidates the sensed data, and hands it over to the C&DH for transmission to the ground. It is also common for the C&DH to hand over all commands related to science instruments to the IC&DH.

Consumables inventory

The spacecraft computer calculates and maintains a table of consumables data, both value and usage rate. This includes available electrical power in the batteries, amount of thruster propellant, and any other renewable or consumable asset. This is periodically telemetered to the ground.

Thermal management

The spacecraft electronics needs to be kept within a certain temperature for proper operation. Generally, the only heat source is the Sun, and the only heat sink is deep space. There are options as to how the spacecraft can be oriented. In close orbit to a planet, the planet may also represent a heat source. Automatic thermal louvers can be used to regulate the spacecraft internal temperature, if they are pointed to deep space. The flight computer's job is to keep the science instrument or communications antennae pointed in the right direction. This might be overridden in case the spacecraft is getting too hot or too cold.

Electrical Power/energy management

The flight Computer needs to know the state-of-charge (SOC) of the batteries at all times, and whether current is flowing into or out of the batteries. It the SOC is getting too low, some operations must be suspended, so the solar panels or spacecraft itself can be re-oriented to maximize charging. In some cases, redundant equipment may be turned off, according to a predetermined load-shedding algorithm. If the spacecraft batteries are fully discharged, it is generally the end of the mission, because pointing to the Sun cannot be achieved, except by lucky accident. Don't bet on it.

Antenna Pointing

The spacecraft communications antennae must be pointed to the large antennae on the ground (Earth) or to a communications relay satellite in a higher orbit (for Earth or Mars). The Antennae can usually be steered in two axis, independently of the spacecraft body. This can be accomplished in the Main flight computer, or be a task for the C&DH.

Safe Hold mode

As a last resort, the spacecraft has a safe-hold or survival mode that operates without computer intervention. This usually seeks to orient the spacecraft with its solar panels to the Sun to maximize power, turn off all non-essential systems, and call for help. This can be implemented in a dedicated digital unit. It used to be the case that the safe-hold mode was implemented in analog circuitry.

Flight Linux

The FlightLinux project in 2002 had the stated goal of providing an on-orbit flight demonstration of the Linux software. The FlightLinux proof-of-concept demonstration was done in conjunction with the on-orbit UoSat-12 mission, from Surrey Space Technology, Ltd. The OMNI project of Code 588 had a breadboard of the Surrey OBC, which was used for testing. This work was funded by the NASA Earth Science Technology Office (ESTO) Advanced Information Systems Technology (AIST) Program. The author was the Principal Investigator.

Because almost all of the effort in developing onboard computer hardware for spacecraft involves adapting existing commercial designs, the logical next step is to adapt COTS software, such as the Linux operating system. Given Linux, many avenues and opportunities become available. Web serving and file transfer become standard features. Onboard LAN and an onboard file system become "givens." Java is trivial to implement. Commonality with ground environments allows rapid migration of algorithms from ground-based to the flight system, and tapping into the world-wide expertise of Linux developments provides a large pool of talent. Full source for the operating system and drivers is available on day one of the project.

Since the one of the goals was keeping the FlightLinux open source we had numerous offers of collaboration on the project. These include representatives of worldwide aerospace companies, and individuals.

The initial FlightLinux software load was approximately 400,000 bytes in size. The nature of the Uosat-12 memory architecture at boot time limited the load size to less than 512,000 bytes. After the loader, which is Read-Only Memory (ROM)-based, completes, 4 megabytes of memory became available. The software load included a routine to setup the environment and a routine to decompress and start the Linux kernel. The kernel is the central portion of the operating system, a monolithic code entry. It controls process management, Input/Output, the file system, and other features. It provides an executive environment to the application programs, independent of the hardware.

Extensive customization of the SETUP routine, written in assembly language, was required. This routine in its original form relied on BIOS (Basic Input/Output System) calls to discover and configure hardware. In the UoSat configuration, there is no BIOS function, so these sections were replaced with the appropriate code. Sections of SSTL code were added to configure the unique hardware of the UoSat computer. The SETUP routine then configured the processor for entry to Protected Mode and invoked the decompression routine for the kernel. The SETUP routine was approximately 750 bytes in length and represented the custom portion of the code for the UoSat software port. This product is usually referred to as the Board Support Package. The remaining code is Commercial-off-the-Shelf (COTS) Linux software. This process would be the same for any FlightLinux port.

The standard Linux SETUP routine, written in assembly language, was modified to be table-driven. This had the added advantage of addressing the export restriction issues.

The breadboard architecture included an asynchronous serial port for debugging. This was used this extensively for debugging the SETUP module. On the spacecraft, the asynchronous port exists, but it is not connected to any additional hardware.

99

FlightLinux was implemented in an incremental manner. The initial software build did a "Hello, World" aliveness indication via the asynchronous port and allows login. The synchronous serial drivers was integrated to allow communication in the flight configuration. The bulk memory device driver, which used the 32-megabyte modules of extended memory as a file system, was added next. The breadboard had a single 32-megabyte module, and there were four modules in the flight configuration. The CAN bus drivers and the network interface were added later.

The FlightLinux Project explored new issues in the use of "free software" and open-source code, in a mission critical application. Open-source code, as an alternative to proprietary software has advantages and disadvantages. The chief advantage is the availability of the source code, with which a competent programming team can develop and debug applications, even those with tricky timing relationships. The Open-Source code available today for Linux supports international and ad hoc standards. The use of a standards-based architecture has been shown to facilitate functional integration. It is a misconception that "free software" is necessarily available for little or no cost. The "free" part refers to the freedom to modify the source code.

A disadvantage of developing with Open Source may be the perception that freely downloadable source code might not be mature or trustworthy. Countering this argument is the growing experience that the Open-Source offerings are as good as, and sometimes better than the equivalent commercial products. What is needed, however, is a strong configuration control mechanism. For the FlightLinux product, the FlightLinux Team assumed the responsibility of making the "official" version available.

Issues on the development and use of Open-Source software on government-funded and mission-critical applications are still to be fully explored.

Given the candidate processors identified in missions under

100

development and planned in the short term, we then examined the feasibility of Linux ports for these architectures. In every case, a Linux port was not only feasible, but is probably available COTS. Each needs to be customized to run on the specific hardware architecture configuration of the target board.

The advantages of Linux are numerous, but the requirements for spacecraft flight software are unique and non-forgiving. Traditional spacecraft onboard software has evolved from being monolithic (without a separable operating system), to using a custom operation system developed from scratch, to using a commercial embedded operating system such as VRTX or VxWorks. None of these approaches have proved ideal. In many cases, the problems involved in the spacecraft environment require access to the source code to debug. This becomes an issue with commercial vendors. Cost is also an issue. When source code is needed for a proprietary operating system, if the manufacturer chooses to release it at all, it is under a very restrictive non-disclosure agreement, and at additional cost. The Linux source is freely available to the team at the beginning of the effort.

As a variation of Linux, and thus Unix, FlightLinux was Open Source, meaning the source code is readily available and free. FlightLinux currently addresses soft real-time requirements and is being extended to address hard real-time requirements for applications such as attitude control. There is a world-wide experience base in writing Linux code that is available to tap.

The use of the FlightLinux operating system simplified several previously difficult areas in spacecraft onboard software. For example, the FlightLinux system imposes a file system on onboard data storage resources. In the best case, Earth-based support personnel and experimenters may network-mount onboard storage resources to their local file systems. The FlightLinux system both provides a path to migrate applications onboard and enforces a commonality between ground-based and space-based resources.

Linux is not by nature or design a real-time operating system. Spacecraft embedded flight software needs a real-time environment in most cases. However, there are shades of real time, specified by upper limits on interrupt response time and interrupt latency. We can generally collect these into hard real-time and soft real-time categories. Examples of hard real-time requirements are those of attitude control, spacecraft clock maintenance, and telemetry formatting. Examples of soft real-time requirements include thermal control, data logging, and bulk memory scrubbing. Many real-time schedulers for Linux are available. These are a Rate Monotonic Scheduler, which treats tasks with a shorter period as tasks with a higher priority, and an Earliest Deadline First (EDF) scheduler. Other approaches are also possible. It is not clear which approach provides the best approach in the spacecraft-operating environment.

A device driver is the low-level software routine that interfaces hardware to the operating system. It abstracts the details of the hardware, in such a way that the operating system can deal with a standardized interface for all devices. In Unix-type operating systems such as Linux, the file system and the I/O devices are treated similarly.

Device drivers are prime candidates for implementation in assembly language, because of the need for bit manipulation and speed. They can also be implemented in higher-order languages such as "c" however. Typical device drivers include those for serial ports, for the mass storage interface, for the LAN interface, etc. Device drivers are both operating system-specific, and specific to the device being interfaced. They are custom code, created to adapt and mediate environments.

The current state of the art of spacecraft secondary storage is bulk memory, essentially large blocks of DRAM. This memory, usually still treated as a sequential access device, is mostly used to hold telemetry during periods when ground contact is precluded. Bulk memory is susceptible to errors on read and write, especially in the space environment, and needs multi-layer protection such as triple-modular redundancy (TMR), horizontal and vertical Cyclic Redundancy Codes (CRC), Error Correcting Codes (ECC), and scrubbing. Scrubbing can be done by hardware or software in the background. The other techniques are usually implemented in hardware. With a Memory Management Unit (MMU), even using 1:1 mapping of virtual to physical addresses, the MMU can be used to re-map around failed sections of memory.

Although we usually think of bulk memory as a secondary storage device with sequential access, it may be implemented as random access memory within the computer's address space. This was the case with UoSat-12.

The Flash File System (FFS) has been developed for Linux to treat collections of flash memory as a disk drive, with an imposed file system. Although we are dealing with DRAM and not flash, we still gained valuable insight from the FFS implementation. In addition, the implementation of Linux support for the personal computer memory card international association (pcmcia) devices provided a useful model.

The onboard computers on the UoSat-12 spacecraft had 128 megabytes of DRAM bulk memory. It is divided into four banks of 32 megabytes each, mapped through a window at the upper end of the processor's address space. This is the specific device driver that the team developed and used as a model for future development of similar modules. The current software of the UoSat-12 onboard computer treats this bulk memory as paged random access memory and applied a scrubbing algorithm to counter radiation induced errors.

The ram disk is a disk-like block device implemented in RAM. This is the correct model for using the bulk memory of the onboard computer as a file system. Multiple RAM disks may be allocated in Linux. The standard Linux utility "mke2fs," which creates a Linux second extended file system, works with RAM disk, and supports redundant arrays of inexpensive disks (RAID) level 0.

This initial version of the driver used memory mirroring, with memory scrubbing techniques applied. In the simplest case, we treated three of the four available 32-megabyte memory pages as a mirrored system. The memory scrubbing technique is derived from the current scheme used by SSTL, as is the paging scheme. The next version of the driver used all four of the available 32-megabyte memory pages with distributed parity. The performance with respect to write speed is expected to be less than with the Level 0, but the memory resilience with respect to errors is expected to be much better.

Given that the Linux operating system is onboard the spacecraft, support for a spacecraft LAN becomes relatively easy. Extending the onboard LAN to other spacecraft units in a constellation also becomes feasible, as does having the spacecraft operate as an Internet node.

For space to ground communication, FlightLinux was planning to use the IP-in-space work validated by GSFC's OMNI (Operating Spacecraft as Nodes on the Internet) Project.

The UoSat-12 configuration allowed for exercising the TCP/IP and CAN bus components of an onboard LAN. Evolving physical layer interfaces for use onboard the spacecraft include 100 megabit Ethernet and Firewire (IEEE-1394).

Having a Linux system enables a plethora of software that runs under Linux. One package is the Beowulf software which implements a cluster of Linux machines. This may use several machines co-located in one satellite, or multiple satellites in a constellation linking their computational resources via inter-satellite communications.

The FlightLinux project ran out of time and funds before it was able to be flown, in part because of ITAR restrictions on the software. It did influence the direction of flight software for later missions.

CFE/cFS

The Core Flight Executive, from the Flight Software Branch, Code 582, at NASA/GSFC, is an open source operating system framework. The executive is a set of mission independent reusable software services and an operating environment. Within this architecture, various mission-specific applications can be hosted. The cFE focuses on the commonality of flight software. The Core Flight System (CFS) supplies libraries and applications. Much flight software legacy went into the concept of the cFE. It has gotten traction within the Goddard community, and is in use on many flight projects, simulators, and test beds (FlatSats) at multiple NASA centers.

The cFE presents a layered architecture, starting with the bootstrap process, and including a real time operating system. At this level, a board support package is needed for the particular hardware in use. Many of these have been developed. At the OS abstraction level, a Platform support package is included. The cFE core comes next, with cFE libraries and specific mission libraries. Ap's habituate the 5th, or upper layer. The cFE strives to provide a platform and project independent run time environment.

The boot process involves software to get things going after power-on, and is contained in non-volatile memory. cFE has boot loaders for the RAD750 (from BAE), the Coldfire, and the Leon3 architecture. The real time operating systems can be any of a number of different open source or proprietary products, VxWorks and RTEMS for example. This layer provides interrupt handling, a scheduler, a file system, and interprocess communication.

The Platform Support Package is an abstraction layer that allows the cFE to run a particular RTOS on a particular hardware platform. There is a PSP for desktop pc's for the cFE. The cFE Core includes a set of re-usable, mission independent services. It presents a standardized application Program Interface (API) to the programmer. A software bus architecture is provided for messaging between applications.

The Event services at the core level provides an interface to send asynchronous messages, telemetry. The cFE also provides time services.

Aps include a Health and Safety Ap with a watchdog. A housekeeping AP for messages with the ground, data storage and file manager aps, a memory checker, a stored command processor, a scheduler, a checksummer, and a memory manager. Aps can be developed and added to the library with ease.

A recent NASA/GSFC Cubesat project uses a FPGA-based system on a chip architecture with Linux and the cFE. CFE and its associated cFS are available as an architecture for Cubesats in general.

The cFE has been released into the World-Wide Open Source community, and has found many applications outside of NASA.

NASA's software Architecture Review Board reviewed the cFE in 2011. They found it a well thought-out product that definitely met a NASA need. It was also seen to have the potential of becoming a dominant flight software architectural framework. The technology was seen to be mature.

The cFS is the core flight software, a series of aps for generally useful tasks onboard the spacecraft. The cFS is a platform and project independent reusable software framework and set of reusable applications. This framework is used as the basis for the flight software for satellite data systems and instruments, but can be used on other embedded systems in general. More information

106

on the cFS can be found at http://cfs.gsfc.nasa.gov

OSAL

The OS Abstraction Layer (OSAL) project is a small software library that isolates the embedded software from the real time operating system. The OSAL provides an Application Program Interface (API) to an abstract real time operating system. This provides a way to develop one set of embedded application code that is independent of the operating system being used. It is a form of middleware.

cFS aps

CFS aps are core Flight System (CFS) applications that are plug-in's to the Core Flight Executive (cFE) component. Some of these are discussed below.

CCSDS File Delivery (CF)

The CF application is used for transmitting and receiving files. To transfer files using CFDP, the CF application must communicate with a CFDP compliant peer. CF sends and receives file information and file-data in Protocol Data Units (PDUs) that are compliant with the CFDP standard protocol defined in the CCSDS 727.0-B-4 Blue Book. The PDUs are transferred to and from the CF application via CCSDS packets on the cFE's software bus middleware.

Limit check (LC)

The LC application monitors telemetry data points in a cFS system and compares the values against predefined red/yellow threshold limits. When a threshold condition is encountered, an event message is issued and a Relative Time Sequence (RTS) command script may be initiated to respond/react to the threshold violation.

Checksum (CS)

The CS application is used for for ensuring the integrity of onboard memory. CS calculates Cyclic Redundancy Checks (CRCs) on the different memory regions and compares the CRC values with a baseline value calculated at system start up. CS has the ability to ensure the integrity of cFE applications, cFE tables, the cFE core, the onboard operating system (OS), onboard EEPROM, as well as, any memory regions ("Memory") specified by the users.

Stored Command (SC)
The SC application allows a system to be autonomously commanded 24 hours a day using sequences of commands that are loaded to SC. Each command has a time tag associated with it, permitting the command to be released for distribution at predetermined times. SC supports both Absolute Time tagged command Sequences (ATSs) as well as multiple Relative Time tagged command Sequences (RTSs).

Scheduler (SCH)
The SCH application provides a method of generating software bus messages at pre-determined timing intervals. This allows the system to operate in a Time Division Multiplexed (TDM) fashion with deterministic behavior. The TDM major frame is defined by the Major Time Synchronization Signal used by the cFE TIME Services (typically 1 Hz). The Minor Frame timing (number of slots executed within each Major Frame) is also configurable.

File Manager (FM)
The FM application provides onboard file system management services by processing ground commands for copying, moving, and renaming files, decompressing files, creating directories, deleting files and directories, providing file and directory informational telemetry messages, and providing open file and directory listings. The FM requires use of the cFS application library.

Data Storage (DS
The DS application is used for storing software bus messages in files. These files are generally stored on a storage device such as a solid state recorder but they could be stored on any file system. Another cFS application such as CFDP (CF) must be used in order

108

to transfer the files created by DS from their onboard storage location to where they will be viewed and processed. DS requires use of the cFS application library.

Memory Manager (MM)
The MM application is used for the loading and dumping system memory. MM provides an operator interface to the memory manipulation functions contained in the PSP (Platform Support Package) and OSAL (Operating System Abstraction Layer) components of the cFS. MM provides the ability to load and dump memory via command parameters, as well as, from files. Supports symbolic addressing. MM requires use of the cFS application library.

Housekeeping (HK)
The HK application is used for building and sending combined telemetry messages (from individual system applications) to the software bus for routing. Combining messages is performed in order to minimize downlink telemetry bandwidth. Combined messages are also useful for organizing certain types of data packets together. HK provides the capability to generate multiple combined packets so that data can be sent at different rates.

Memory Dwell (MD)
The MD application monitors memory addresses accessed by the CPU. This task is used for both debugging and monitoring unanticipated telemetry that had not been previously defined in the system prior to deployment. The MD application requires use of the cFS application library.

Software Bus Network (SBN)
The SBN application extends the cFE Software Bus (SB) publish/subscribe messaging service across partitions, processes, processors, and networks. The SBN is prototype code and requires a patch to the cFE Software Bus code. This is now included in the software library.

<u>Health and Safety (HS)</u>

The HS application provides functionality for Application Monitoring, Event Monitoring, Hardware Watchdog Servicing, Execution Counter Reporting (optional), and CPU Aliveness Indication (via UART).

Being open source, you can write your own cFS aps for specific applications, or modify existing ones. However, you should submit them back to the owner (NASA-GSFC) for review and validation so they become a part of the official package.

Flight Software Complexity

There was a 2009 study of the Complexity of Flight Software with NASA Headquarters and most of the Field Centers participating. The sponsor was the NASA Office of the Chief Engineer. There charter was to "Bring forward deployable technical and managerial strategies to effectively address risks from growth in size and complexity of flight software." The report is now dated, and needs updating.

The first issue addressed was that of growth in flight software size. They had plotted mission software size in terms of lines of code per year of mission, and gotten an exponential growth curve, with a 10x growth every 10 years, from 1968 to 2004. they had seen similar growth curves in Defense Systems, aircraft, and automobiles.

Software size, in terms of lines of code, is an indicator of complexity. Not a great indicator, but certainly one that can be measured.

Then, they set off to define complexity of software. This involves not only the number of components of a system, bu also their inter-relationships. This leads to that fact that a systems has a certain essential complexity, which comes from the problem being addressed. There is also extraneous or incidental complexity, that gets added because of the solution chosen. Essential complexity comes from the problem domain and the requirements. The only way to reduce it is to downscope the problem. It can be moved later to operations, but not erased.

One of the major finding was that "Engineers and scientists often don't realize the downstream complexity entailed by their decisions." It was also noted that "...NPR 7123, NASA System Engineering Requirements, specifies in an appendix of "best typical practices" that requirements include rationale, but offers no guidance on how to write a good rationale or check it."

One good recommendation was that Software Architecture was a little-known or not-well understood element of software design, but an essential one. Another finding, in the NASA context, was that often a specific optimized design vastly increases operational complexity. Incidental complexity, though, comes from design choices.
They found that COTS software was a mixed blessing, in that it comes with features not needed. Although not needed, these features require additional testing, and increase the complexity. And, it is more complex to understand and remove them, then to test them.

One of the "take-away" messages was that flight software is increasing in complexity because we are solving increasing complex problems. One solution is to address complexity with architecture.

They quote a 1968 NATO report with the same concerns for the same reasons, although they considered 10,000 lines of code as complex, then.

NASA recommended more emphasis on Fault Detection and containment.

They defined these characteristics for Flight Software:

"No direct user interfaces such as monitor and keyboard. All interactions are through uplink and downlink.

Interfaces with numerous flight hardware devices such as thrusters, reaction wheels, star trackers, motors, science instruments, temperature sensors, etc.

Executes on radiation-hardened processors and microcontrollers that are relatively slow and memory-limited. (Big source of incidental complexity)

Performs real-time processing. Must satisfy numerous timing constraints (timed commands, periodic deadlines, async event response). Being late = being wrong."

An interesting chart derived for JPL Missions (planetary) shows a vertical axis of software size times processor speed (bytes, mips) and a horizontal axis of time, where the curve through various missions is linear; ie, exponential growth, with a doubling time under two years.

The study pointed out that each step of the lifecycle process, requirements, design, coding and testing, both removed defects, and inserted new one. Thus, there are residual defects that ship with the system. Some of these are never found.

We can focus on reducing the defect insertion rate, or increasing the removal rate, but the bad news is, we'll never drive the rate to zero. Thus, there will be residual defects at launch. From empirical evidence, 1 million lines of code will have 900 benign defects, 90 medium level, and 9 potentially fatal. (Wait, did we say a card had 100 million lines of code?) This is based on a count of 1 residual defect per 1000 lines of code, an across-industries average for embedded code. What this leads us to conclude is that there is a current upper limit to system software complexity, measured in lines of code, because, beyond a certain size, the probability of mission failure tends to 1.

Architecture of a embedded or flight system is an essential part of the development process. Architecture tells us what we are building, not necessarily how. The architecture phase of system engineering has been slow to be adopted. The principles noted are that

- "Architecture" is an abstraction of a system that suppresses some details.
- Architecture is concerned with the public interfaces of elements and how they interact at runtime.
- Systems comprise more than one structure, e.g., runtime processes, synchronization relations, work breakdown, etc. No single structure is adequate.
- Every software system has an architecture, whether or not documented, hence the importance of architecture documentation.
- The externally visible behavior of each element is part of the architecture, but not the internal implementation details.
- The definition is indifferent as to whether the architecture is good or bad, hence the importance of architecture evaluation."

As things get more and more complex, even everyday things, we need to develop better ways to develop and verify software, whether it flies in space, or runs on our phone.

Technology Readiness Levels

The Technology readiness level (TRL) is a measure of a device's maturity for use. There are different TRL definitions by different agencies (NASA, DoD, ESA, FAA, DOE, etc). TRL are based on a scale from 1 to 9 with 9 being the most mature technology. The use of TRLs enables consistent, uniform, discussions of technical maturity across different types of technology. We will discuss the NASA one here, which was the original definition from the 1980's. TRL's apply to software as well as hardware.

Technology readiness levels in the National Aeronautics and Space Administration (NASA)

The TRL assessment allows us to consider the readiness and risk of our technology elements, and of the system.

1. Basic principles observed and reported
This is the lowest "level" of technology maturation. At this level, scientific research begins to be translated into applied research and development.

2. Technology concept and/or application formulated
Once basic physical principles are observed, then at the next level of maturation, practical applications of those characteristics can be 'invented' or identified. At this level, the application is still speculative: there is not experimental proof or detailed analysis to support the conjecture.

3. Analytical and experimental critical function and/or characteristic proof of concept.

At this step in the maturation process, active research and development (R&D) is initiated. This must include both analytical studies to set the technology into an appropriate context and laboratory-based studies to physically validate that the analytical predictions are correct. These studies and experiments should constitute "proof-of-concept" validation of the applications/concepts formulated at TRL 2.

4. Component and/or breadboard validation in laboratory environment.

Following successful "proof-of-concept" work, basic technological

114

elements must be integrated to establish that the "pieces" will work together to achieve concept-enabling levels of performance for a component and/or breadboard. This validation must be devised to support the concept that was formulated earlier, and should also be consistent with the requirements of potential system applications. The validation is "low-fidelity" compared to the eventual system: it could be composed of ad hoc discrete components in a laboratory

TRL's can be applied to hardware or software, components, boxes, subsystems, or systems. Ultimately, we want the TRL level for the entire systems to be consistent with our flight requirements. Some components may have higher levels than needed.

5. Component and/or breadboard validation in relevant environment.

At this level, the fidelity of the component and/or breadboard being tested has to increase significantly. The basic technological elements must be integrated with reasonably realistic supporting elements so that the total applications (component-level, sub-system level, or system-level) can be tested in a 'simulated' or somewhat realistic environment.

6. System/subsystem model or prototype demonstration in a relevant environment (ground or space).

A major step in the level of fidelity of the technology demonstration follows the completion of TRL 5. At TRL 6, a representative model or prototype system or system - which would go well beyond ad hoc, 'patch-cord' or discrete component level breadboarding - would be tested in a relevant environment. At this level, if the only 'relevant environment' is the environment of space, then the model/prototype must be demonstrated in space.

7. System prototype demonstration in a space environment.

TRL 7 is a significant step beyond TRL 6, requiring an actual

115

system prototype demonstration in a space environment. The prototype should be near or at the scale of the planned operational system and the demonstration must take place in space.

The TRL assessment allows us to consider the readiness and risk of our technology elements, and of the system.

New Paradigms and Technology

Spacecraft computer systems followed the trend from purpose-built custom units to those based on standard microprocessors. However, the space environment is very unforgiving in many areas, the chief one being radiation. Commercial electronic parts do not last very long in orbit.

So, we have seen control systems for spacecraft go from hardwired logic to a general purpose CPU architecture programmed with software. ASIC's, or Application Specific Integrated Circuits, are also produced in radiation hard versions. This is sort of a throwback to the hardwired approach, but has its advantages. The next step is systems built from FPGA's, or Field Programmable Gate Arrays. Here, the hardware architecture itself is programmed.

With an FPGA, if you need 18 bits, more than 16 but less than 32, you can easily build an 18-bit machine. You can include the instructions and data paths you need, and leave out the others. It is a way to build a hardwired unit by programming the hardware into the right architecture. Library's of components are available to instantiate various high-level functions (such as a cpu or decoder) into the FPGA. The available complexity of radiation-hard parts removes device capability as a limitation. What remains is the ability to design, implement, test, and verify the parts – a significant challenge. Because FPGA's are not just hardware and not just software, there is a steep learning curve and some conceptual blockbusting needed for software or hardware engineers to master this new paradigm.

The next big step is to utilize the fact that FPGA's are reprogrammable. This brings in many of the same issues that were faced by the earliest software based flight systems. Configuration Management is key.

Let's take a look at the Cost of Change for onboard systems before and after launch: Launch is a particularly significant event – the spacecraft comes out of the lab and is not hands-on anymore.

Implementation	cost to change before launch	change after launch
Hardwired	$$	no
CPU + Software	$$	yes
ASIC	$$$	no
FPGA	$	no
Reprog. FPGA	$	yes

A hardwired system is expensive to change before launch, and impossible to change afterwards. With software, cost is still high, but the system can be changed (reprogrammed) after launch. With an ASIC, we have perhaps a smaller implementation, a system-on-a-chip, but the development costs are large, and it is not possible to change the configuration after launch. An FPGA-based system is similar, but the costs are lower. And, a reprogrammable FPGA provides the ability to modify the flight system after launch.

Let's now look at the complexity in terms of numbers of logic gates of some flight units, from the Apollo Guidance computer to Recently announced radiation-hard reprogrammable FPGA's. The latest unit from Xilinx does not list its complexity in terms of gate-equivalents, because Xilinx claims that is a meaningless measure.

Flight Computational Units

117

Unit	gates	rad-hard
Apollo GC	5000	no
NSSC-1	6000	yes
ASIC	10M	no
ASIC-RH	500k	yes
Actel 1280	8000	no
RT-SX	50-60k	yes
RT-AX	4M	yes
Xilinx V-4	1M	mostly

SpaceCube

The Spacecube is a Goddard Space Flight Center family of high performance reconfigurable computers for space flight use. Version 1, which used Xilinx Virtex-4 FPGA's, has flown in space. The device features a small footprint, 4 by 4 by 3 inches, and is low power. The processing engines are four PPC405 @ 450MHz within the FPGA. It also features configurable I/O: LVDS, SpaceWire, or RS422. It is a flexible and stackable Architecture. The first flight of SpaceCube was on the Hubble Servicing Mission-4. The 2nd flight was on STS-129, where the unit was attached to the outside of the International Space Station. The original unit features four PowerPC cpu's implemented within FPGA's, and operating at 450 MHz. Memory scrubbing is used to mitigation radiation hits to memory. The architecture supports VxWorks, Linux, or RTEMS. A pair of Aeroflex UT6325 RISC microcontroller modules implement the voting logic. Each PowerPC has access to its own 128 megabytes of static ram. The SpaceCube 4 inch x 4 inch board is designed with a stacking connector, so additional boards for I/O can be added as needed. The stack implements 8 high speed and two low speed busses. The onboard reconfigurable part includes dual 56,860 logic cells, and dual 4,176 kbytes of ram. There are 73 user-defined I/O lines.

Newer models of the Spacecube have been developed using the

newer radiation-hard parts, and with expanded computational and interface capabilities. An intermediate version of Spacecube, 1.5, upgraded to Xilinx commercial grade Virtex-5 FPGA's. These were tested on sounding rockets. Version 2.0 uses the inherently radiation hard Virtex-5.

Spacecube 2.0 uses a standard 3U form factor, 4" x 6". It presents cpu, fpga, and dsp computational capability. It has a high 500 mips/watt factor, with lower cost. It is radiation tolerant, not radiation hard.

The processor card includes dual FPGA's, each with dual PowerPC 400's, 512 megabytes of ram, and 2 gigabytes of flash. A third modules includes a rad-hard V5 SIRFVirtex-5QV FX130 SIRF with 8 megabytes of rad-hard SRAM, 64 megabytes of ROM, 8 gigabytes of flash, and 512 megabytes of SDRAM. The SIRF was developed by the Air Force Research Lab, and does not include the embedded PowerPC's.

Most spacecraft flying today have less computing power than the cheapest phone on the market. We can no longer afford this scenario. Spacecraft computing needs to advance to allow continuation of our vision of space exploration and to understand and protect Earth.

Reconfigurable computing is the solution. The *Radiation Hard By Design* FPGA enables a new paradigm for Space Computing. Here, Non-read-hard parts are used with voting logic and other techniques to gain access to the commercial state-of-the-art performance and capacities, with the use of a minimal number of rad-hard parts (the voter). It has been shown to be a good approach.

Next Generation space computers

In 2013, The Air Force Research Lab, Space Vehicles Directorate, in conjunction with NASA, issued a pre-solicitation for the Next Generation Space Processor. This is considered an R&D effort, resulting in hardware that would be viable through the year 2030.

The rad-hard processor is to be based on a commercial architecture, able to implement parallel processing. It is to have a minimum of 24 cores, with a global terabyte of memory space, and include floating point. The performance goal is 24 giga-operations per second and 10 giga floating point operations per second, while drawing 7 watts or less. It has to have hardware support for real-time, and for power control. It has to have integral fault detection and correction. It needs eight 10 gigabit-per-second serial I/O ports. It is to support Linux. And c/c++. Boeing was selected to implement this program.

At the same time, the European Space Agency, ESA, is seeking a Next-Generation Digital Signal Processor for Space Applications.

Exploring our Solar System

This section discusses the flight computers for missions to the Sun and other planets of our solar system. Each represents unique challenges. The Moon was the first extra-terrestial body to be explored. It is close enough that the communication time is about ½ second, and lunar spacecraft could be controlled from Earth. But, the lack of communication with the craft when they are behind the moon, from the Earth viewpoint, dictated at least a stored command capability. Rovers on the face of the moon towards Earth are in continuous contact.

Exploring the Sun

We get all our energy from the huge thermal reactor some 8 light-minutes away. As you get closer to the Sun, its get hotter, and there is more radiation in terms of energetic particles.

The 1980 Solar Maximum Mission observed the Sun in the spectrum of gamma rays, X-rays, and Ultraviolet. SMM had a failure in its electronics months after launch, but was repaired by a subsequent Shuttle mission. SMM used the NASA Standard Spacecraft Computer (NSSC-1) constructed of discrete logic elements. The author had flight software onboard SMM, when it reentered the atmosphere and burned in 1989.

The Solar Terrestial Relations Observatory (Stereo) is a dual spacecraft mission to observe the Sun, launched in 2006. One is ahead of the Earth in orbit, the other behind. This gives three points of view of solar phenomena. Stereo's onboard computer uses the dual redundant Integrated Electronics Module (IEM) which has a RAD6000 cpu, as well as Actel FPGA's with soft-core P24 and CPU24 architectures. The P24 architecture is a 24-bit minimal instruction set computer (MISC). It is still operating, as of this writing.

Exploring Mercury

The U. S. Messenger (Mercury Surface, Geochemestry and Ranging) mission to Mercury, the closest planet to the Sun, was launched in 2004. It is currently orbiting the hottest planet. It's Integrated Electronics Module has two Rad-Hard RAD6000 processors. One is the main, the other is a fault protection processor, operating at a slower clock rate (10 MHz vs 25 MHz). The modules are also duplicated. Two solid stage data recorders with 1 gigabyte of storage capacity each are used. No landing on Mercury has been attempted, although it would be feasible in the *twilight zone* between the extremely hot solar facing side, and the much colder space facing side. The mission was completed in 2015.

Exploring Venus

The Soviet space program sent a series of probes to Venus. Early efforts were either crushed in the dense atmosphere, or suffered thermal damage. The surface temperature is high enough to melt some metals. This is very hard on computers, and electronics in general. On its way to Saturn, via Jupiter, the Cassini spacecraft imaged our neighbor planet in 1988 and 1999.

The Pioneer Venus spacecraft was launched into Venus orbit in 1978, and returned data until 1992. It did not use a computer, but an attitude controller built from discrete components. The Venus Express, a ESA mission, in 2005 used multiple Xilinx FPGA's.

Exploring the Moon

Early in the era of space exploration, a series of rover vehicles were sent to the Earth's moon. These were designed as precursors to a crewed visit. From the mid-1960's through 1976, there were some 65 unmanned landings on the moon. The moon is still the subject of intense study, with missions from the United States, Russia, China, India, the European Union, and Japan.

The Soviet Union launched a series of successful lunar landers, sample return missions, and lunar rovers. The Lunokhod missions, from 1969 through 1977, put a series of remotely controlled vehicles on the lunar surface.

The latest mission to study the Moon is the Lunar Reconnaissance Orbiter, LRO, from NASA/GSFC. It was launched in 2009, and is still operating. It is is a polar orbit, coming as close as 19 miles to the lunar surface. It is collecting the data to construct a highly detailed 3-D map of the surface. Up to 450 Gigabits of data per day are returned to Goddard.

The LRO uses the RAD750 processor, on a CompactPCI 6U circuit card. The card provides a 1553 bus interface, and a 4-port Spacewire router. The cpu has 36 Megabytes of rad-hard sram, 4 megabytes of EEProm, and a 64k ROM. The SpaceWire functionality is provided by an ASIC chip, with access to its own 8 megabytes of SRAM. The transport layer of the Spacewire protocol is implemented in the ASIC in hardware. The 1553 interface is implemented in an FPGA

The cpu operates with a 132 MHz clock, and the backplane bus runs at 66 MHz. It consumes between 5 and 19 watts of power, and weighs around 3.5 pounds.

The cpu communicates to other spacecraft electronics over the backplane cPCI bus, the 1553, or the Spacewire. The High data volume camera uses the SpaceWire interface at up to 280 MHz, and other instruments use the 1553. The onboard data storage for data is a 400 gigabit mass memory unit, using SDRAM.

Mars Missions

The Viking program was a pair of spacecraft sent to Mars in 1975. Each spacecraft consisted of an orbiter, and a lander. The Viking landers used a Guidance, Control and Sequencing Computer (GCSC) consisting of two Honeywell HDC 402 24-bit computers with 18K of plated-wire memory, while the Viking orbiters used a Command Computer Subsystem (CCS) with two custom-designed 18-bit bit-serial processors. There were 6 bits for the opcode, and 12 bits for the data address. The data format was 2's complement. The machine supported 32 interrupts. It was programmed in assembly language. The Honeywell machine had 47 instructions, and used two's complement representation for data. The architecture may have been derived from the Honeywell 316/516 models.

Mars Rover Sojourner

The Mars Pathfinder mission landed on Mars on July 4, 1997. It carried a Rover named Sojourner, which was a 6-wheeled design, with a solar panel for power, but the batteries were not rechargeable. The rest of the lander served as a base station. Communication with the rover was lost in September. The Rover used a single Intel 80C85 8-bit CPU with a 2 MHz clock, 64k of ram, 16 k of PROM, 176k of non-volatile storage, and 512 kbytes of temporary data storage. It communicated with Earth via the base station using a 9600 baud UHF radio modem. The communication loss leading to end of mission was in the base station communication, while the Rover remained functional. The Rover had three cameras, and an x-ray spectrometer.

The computer in the mission base station on Mars was a single RS-6000 CPU, with 1553 and VMEbuses. The software was the VxWorks operating system, with application code in the c language. The base station computer experienced a series of resets on the Martian surface, which lead to an interesting remote debugging scenario.

The operating system implemented pre-emptive priority thread (of execution) scheduling. The watchdog timer caught the failure of a task to run to completion, and caused the reset. This was a sequence of tasks not exercised during testing. The problem was debugged from Earth, and a correction uploaded.

The cause was identified as a failure of one task to complete its execution before the other task started. The reaction to this was to reset the computer. This reset reinitialized all of the hardware and software. It also terminates the execution of the current ground commanded activities.

The failure turned out to be a case of priority inversion. The higher priority task was blocked by a much lower priority task that was holding a shared resource. The lower priority task had acquired this resource and then been preempted by several medium priority tasks. When the higher priority task was activated, it detected that the lower priority task had not completed its execution. The resource that caused this problem was a mutual exclusion semaphore used to control access to the list of file descriptors that the select() mechanism was to wait on.

The Select mechanism creates a mutex (mutual exclusion mechanism) to protect the "wait list" of file descriptors for certain devices. The vxWorks pipe() mechanism is such a device and the Interprocess Communications Mechanism (IPC) used was based on using pipes. The lower priority task had called Select, which called other tasks that were in the process of setting the mutex semaphore. The lower priority task was preempted and the operation was never completed. Several medium priority tasks ran until the higher priority task was activated. The low priority task attempted to send the newest high priority data via the IPC mechanism which called a write routine. The write routine blocked, taking control of the mutex semaphore. More of the medium priority tasks ran, still not allowing the high priority task to run, until the low priority task was awakened. At that point, the scheduling task determined that the low priority task had not completed its cycle (a hard deadline in the system) and declared the error that initiated the reset. The reset had the effect of wiping out most of the data that could show what was going on. This behavior was not seen during testing. It was successfully debugged and corrected remotely by the JPL team.

References
http://www.nasa.gov/mission_pages/mars-pathfinder/
http://research.microsoft.com/en-us/um/people/mbj/Mars_Pathfinder/

MER – Mars Exploration Rovers *Spirit & Opportunity*

The MER are six-wheeled, 400 pound solar-powered robots, launched in 2003 as part of NASA's ongoing Mars Exploration Program. *Opportunity* (MER-B) landed successfully at Meridiani Planum on Mars on January 25, 2004, three weeks after its twin *Spirit* (MER-A) had landed on the other side of the planet. Both used parachutes, a retro-rocket, and a large airbag to land successfully, after transitioning the thin atmosphere of Mars.

For power, they use 140 watt solar arrays and Li-ion batteries. The Rovers require 100 watts for driving, One problem that was noted was that the Martian dust storms cover the solar panels with fine dust, reducing their efficiency. This resulted in the use of a radioisotope generator on a subsequent mission. It's been observed that Rovers often use more energy in path planning, than to execute the actual path.

The onboard computer uses a 20 MHz RAD6000 CPU with 128 MB of DRAM, 3 MB of EEPROM, and 256 MB of flash memory on a VME bus. There is a 3-axis inertial measurement unit, and nine cameras The Rovers communicate with Earth via a relay satellite in Mars orbit, the Mars Global Surveyor spacecraft. They also have the ability to communicate directly, at a lower data rate.

The lander and rover include Xilinx FPGA's for mission critical functions. The rover's motor controllers are implemented in the FPGA's. These operated through the worst Solar Flare ever measured. These are particularly bad at Mars, since it has no discernible magnetic field, or the equivalent of the Earth's Van Allen Belts. The FPGA's have SEU detection and use scrubbing for error correction.

The Spirit unit became stuck in 2009, and engineers were unable to free it after 9 months of trying. It was re-tasked as a stationary sensor platform. Contact was lost in 2010.

This is an ongoing mission. It was originally planned for 90 days, but the *Opportunity* Rover is still collecting useful data regarding potential life on our sister planet some 11 years later as of this writing. It has traveled over 35 kilometers on the Martian surface. Ground based test units are used at JPL for evaluating problems seen on Mars, and for evaluating software and procedural fixes.

Mars Science Laboratory Curiosity

The Mars Science Laboratory landed successfully on the Martian surface on August 6, 2012. It had been launched on November 26, 2011. It's location on Mars is the Gale crater, and was a project of NASA's Jet Propulsion Laboratory. The project cost was around $2.5 billion. It is designed to operate for two Martian years (sols). The mission is primarily to determine if Mars could have supported life in the past, which is linked to the presence of liquid water.

The Rover vehicle *Curiosity* weights just about 1 ton (2,000 lbs.) and is 10 feet long. It has autonomous navigation over the surface, and is expected to cover about 12 miles over the life of the mission. The platform uses six wheels The Rover Compute Elements are based on the BAE Systems' RAD-750 CPU, rated at 400 mips. Each computer has 256k of EEprom, 256 Mbytes of DRAM, and 2 Gbytes of flash memory. The power source for the rover is a radioisotope thermal power system providing both electricity and heat. It is rated at 125 electrical watts, and 2,000 thermal watts, at the beginning of the mission. The operating system is WindRiver's VxWorks real-time operating system.

The computers interface with an inertial measurement unit (IMU) to provide navigation updates. The computers also monitor and control the system temperature. All of the instrument control, camera systems, and driving operations are under control of the onboard computers.

Communication with Earth uses a direct X-band link, and a UHF link to a relay spacecraft in Mars orbit. At landing, the one-way communications time to Earth was 13 minutes, 46 seconds. This varies considerably, with the relative positions of Earth and Mars in their orbits around the Sun. At certain times, when they are on opposite sides of the Sun, communication is impossible.

The science payload includes a series of cameras, including one on a robotic arm, a laser-induced laser spectroscopy instrument, an X-ray spectrometer, and x-ray diffraction/fluorescence instrument, a mass spectrometer, a gas chromotograph, and a laser spectrometer. In addition, the rover hosts a weather station, and radiation detectors. There is cooperation between in-space assets and ground rovers in sighting dust storms by the meteorological satellite in Mars orbit.

In 2013, NASA uploaded a software upgrade to Curiosity's operating System. Overall, it took a week to install.

References

http://www.nasa.gov/mars

www.nasa.gov/msl/

http://en.wikipedia.org/wiki/Mars_Science_Laboratory

www.space.com/16385-curiosity-rover-mars-science-laboratory.html

http://www.windriver.com/announces/curiosity/Wind-River_NASA_0812.pdf

Maven

NASA's Maven mission to Mars is an orbiter. It is studying the Martian atmosphere It was launched in November of 2013, and reached Mars in September of 2014.

MAVEN is equipped with a RAD-750 Central Processing Board manufactured by BAE Systems in Manassas, Va. The processor can endure radiation doses that are a million times more extreme than what is considered fatal to humans. The RAD750 CPU itself can tolerate 200,000 to 1,000,000 rads. Also, RAD750 will not suffer more than one event requiring interventions from Earth over a 15-year period.

The RAD-750 was released in 2001 and made its first launch in 2005 aboard the Deep Impact Spacecraft. The CPU has 10.4 million transistors. The RAD750 processors operate at up to 200 megahertz, processing at 400 MIPS. The CPU has an L1 cache memory of 2 x 32KB (instruction + data) - to improve performance, multiple 1MB L2 cache modules can be implemented depending on mission requirements.

The RAD750 operates at temperatures of -55°C to 125°C with a power consumption of 10 Watts.

Asteroid exploration

The Pioneer-11 spacecraft was sent to study the far reaches of the solar system It passed through the Asteroid belt on its way to Jupiter and Saturn. It used a custom design, TTL computer in 1973.

There are thousands of asteroids, and it seems there may be as many types. This means that exploring the known asteroids is a daunting challenge. On the other hand, the asteroids can be a significant source of raw materials for Earth. But, a conventional survey and exploration approach would take too long. What is needed instead is a multitude of autonomous and flexible nano-spacecraft. The architectural model is a swarm (social insect model) distributed intelligence. The platform of low cost, low power, low weight could involve a nano-spacecraft with solar sails. The computation power of the individual nano-spacecraft can be combined into a "cluster of convenience" to address computationally challenging problems as they emerge, and on the spot. This might be a good opportunity for a Cubesat-influenced architecture.

Comets

The Deep Impact mission returned images of the surface of comet

Borrelly in 2001. That surface was hot (26-70C) , dry, and dark. In July of 2005, the same mission sent a probe into Comet Tempel 1. It created a crater, allowing imaging of subsurface material. Water ice was seen. Comet Borrely has a coma, which proved to be vaporized subsurface water ice. Deep Impact went on to complete a flyby of Comet Hartley-2 in 2010. The impactor used a RAD-750 flight computer with an imaging sensor, an inertial measurement unit, and four thrusters. It also included a star tracker, and an S-band communications link.

The 1999 Stardust mission retrieved sample material from the tail of Comet Wild 2 and returned it to Earth in 2006. The spacecraft computer was a RAD6000 with 128 megabytes of memory. The flight software took up 20% of the memory space, allowing for storing data when not in contact with Earth. The real time operating system was VxWorks. The comet material was captured in aerogel. Subsequently, 7 particles of interstellar dust were found in the aerogel. The mission also imaged the asteroids Annefrank, and was redirected to Tempel1 after the primary mission was complete. The RAD6000 cpu was used in the Command & Data Handling subsystem of Stardust.

ESA's Rosetta probe is in orbit around the comet Churyumov-Gerasimenko. It released a lander, Philae, which successfully touched down on the comet's surface in 2014. The onboard system has 3 gigabytes of solid state memory. The lander used a Harris RTX2020 processor. The lander communicated with the main spacecraft over a 32kbps link.

The memory size for the main processor was 1MWord RAM and 512 KWords EEPROM for each of 4 processors, and 512KWords PROM (redundant) accessible from each processor.

The Gas Giants

The Gas giants are the planets Jupiter, Saturn, Uranus, and

Neptune. These are the responsibility of the Jet Propulsion Laboratory, under contract to NASA. The RCA CMOS 1802 8-bit unit was used on JPL's Voyager, Viking, and Galileo space probes. Prior to the Voyager mission, JPL was using simple sequencers purpose-built, and not based on a microprocessor architecture. This Command Computer System (CCS) architecture was 18-bit.

Voyager had a computer command subsystem (CCS) that controlled the imaging cameras. The CCS was based on an earlier design used for the Viking mission. The Attitude and Articulation control system (AACS) controlled the orientation of the spacecraft, and the movement of the camera platform. It is essentially the same computer as the CCS. The Data computer was constructed form a custom 4-bit cmos component.

The ESA/NASA Ulysses mission visited the Jupiter system in 1992 and 2000, and collected data on the magnetosphere. This was a swing-by mission, as Ulysses was primarily to observe the Solar poles. Ulysses used a 280w RTG for power as it swung far from the Sun. Ulysses did not have a flight computer, but used a Central Terminal Unit (CTU) and tape recorders for data storage. Each had a capacity of 45 megabits. The unit had an Attitude and Orbital Control System (AOCS), a purpose built unit, dual redundant, weighting 100 kilograms.

Cassini observed the planet from close-up in the year 2000, and studied the atmosphere. It used RTG's for power, and MIL-STD-1750A control computers.

Galileo entered Jupiter orbit in 1995, and returned data on the planet and the four Galilean moons until 2003. Three of the moons have thin atmospheres, and may have liquid water. The moon Ganymede has a magnetic field. Galileo was in the right place at the right time to see the comet Showmaker-Levy-9 enter the Jovian atmosphere, and launched an atmospheric probe. Galileo used six of the 8-bit RCA 1802 microprocessors, operating at 1.6 MHz.

These units had been fabricated on a sapphire substrate for radiation hardness. The attitude and Articulation Control System used two bit-slice machines built from the AMD2901 chips.

The Juno mission to Jupiter used the RAD-750 flight computer. The spacecraft enter jovian polar orbit in July, 2016. It was a highly elliptical orbit, ranging from2,600 miles to 8.1 million miles. The mission lifetime is set by the exposure to Jupiter's trapped radiation belts. At the end of mission, the spacecraft entered the Jovian atmosphere and burned.

Saturn

Cassini was the fourth spacecraft to study Saturn., which has rings, although smaller than Jupiter. The rings were confirmed by the Voyager spacecraft in the 1980's. Cassini entered into Saturnian orbit, and is still returning data.. The one-way communications time varies form 68-84 minutes. It has also collected data on the Saturnian moons Titan, Enceladus, Mimas, Tethys, Dione, Rhea, Iapetus, and Helene . Things are strange in the Saturnian system. Cassini observed a hurricane in 2006 on the planet's south pole. It appears to be stationary at the pole, 5,000 miles across, 40 miles high, with winds of 350 mph. The large moon Titan has lakes of a liquid hydrocarbon, with possible seas of methane and ethane. Cassini launched a probe *Huygens* to Titan, and it landed on solid ground below the atmosphere. Huygens used a 1750A fight computer The Cassini mission was responsible for the discovery of seven new moons of Saturn.

Pluto and Beyond

The New Horizons Mission spacecraft carries two computer systems, the Command and Data Handling system and the Guidance and Control processor. Each of the two systems is duplicated for redundancy, for a total of four computers. The processor used is the Mongoose-V, a 12 MHz radiation-hardened version of the MIPS R3000 CPU. Multiple clocks and timing routines are implemented in hardware and software to help prevent faults and downtime.

To conserve heat and mass, spacecraft and instrument electronics are housed together in IEMs (Integrated Electronics Modules). There are two redundant IEMs. Including other functions such as instrument and radio electronics, each IEM contains computers.

In March of 2007, the Command and Data Handling computer experienced an uncorrectable memory error and rebooted itself, causing the spacecraft to go into safe mode. The craft fully recovered within two days, with some data loss on Jupiter's magnetotail. Significant information on Pluto starting to be received in 2015.

New Horizons completed its primary mission at Pluto, and is going further out in space to do a flyby of a Kuiper belt object in 2019. There is a mission extension through 2021.

Commercial Ventures

It is not just governments who are funding space exploration. There is money to be made, and commercial ventures are entering the field. Space in unforgiving, and there is a steep learning curve. Nevertheless, progress is being made in the commercial arena. A lot of the flight computer architecture is proprietary, but there is some information available.

SpaceX

SpaceX started out as Space Exploration Technologies Corporation in 2002. It is based in California, and was founded by Elon Musk, of PayPal and Tesla Motors fame. It employs over 4,000.

SpaceX is a big proponent of Linux code for both flight and

support environments. One of their big selling points in general is re-usability of hardware and software. They tend to build what they need in-house. Some of the technology used in the Tesla electric vehicle is applicable. Space-X has a general policy of using triple redundant systems.

The Falcon launch vehicle uses nine of the SpaceX Merlin engines, for a total thrust of over 1 millions pounds-force. The second stage of the vehicle uses a single Merlin engine. The vehicle avionics is single-fault tolerant, and is being human-rated. It uses triple dual-core x86 architecture computers, running linux. The computers interface with a series of device controllers, GPS, an inertial platform, and command and telemetry via c-band radio link. A FalconHeavy vehicle uses three Falcon rockets in a configuration similar to the Titan-III. It has a lift capability of 53 metric tons to orbit. There are thirty cpus for the engine control, running linux.

The reusable Dragon capsule is being developed for crewed spaceflight. At the moment, it is proving itself with unmanned supply missions to the International Space Station. NASA has contracted for a dozen flights. The capsule is capable of fully autonomously rendezvous and docking. It's cargo capacity is 6,000 kilograms. With 10 cubit meters of pressurized payload volume, and 14 cubic meters unpressurized. It includes three dual fault-tolerant flight computers. It is designed to land upright on a platform via parachute. The Dragon capsule's computers run predominately c++ code on linux. The Dragon capsule displays use a modified Tesla touchscreen unit, which includes nVidia Tegra system-on-a-chip units. Tesla hires a lot of top software engineering talent from the gaming industry.

Amsat

Radio Amateurs (HAM operators) have been involved with satellites since the earliest days, building their own small communications satellites, and hitching a ride with other missions. These are spacecraft built in basements and garages, but usually by

134

engineers who's day job is the same thing. Students are active participants in the project, which kicked off around 1969.

OSCAR, the Orbiting Satellite Carrying Amateur Radio was a successful program to provide a repeater in space for the HAM community.

Most of these units used the RCA 1802 8-bit processor, because of its low cost and inherent radiation tolerance. It is not the easiest microprocessor to program, though.

The homebuilt spacecraft are sophisticated units, that sense the Earth & the Sun, and control their attitude and pointing with electromagnets that react against the Earth's field.

Amateur spacecraft are operating in orbit, and projects are in place to put amateur satellites at Geosynchronous orbit, and for planetary missions. These have mostly migrated to the Cubesa architecture.

Small Satellites

Satellites have gotten smaller and less expensive, opening space exploration to Universities, small companies, and even individuals. Smaller satellites can be launched as secondary payloads with large satellites, in many cases replacing the ballast that would be used to adjust the weight and trim. Also, multiple small satellites can share a launch vehicle, if they going to the same orbit. These small satellites need the same overall complement of equipment as the bigger models, including a flight computer. They need attitude control (or, at least, sensing), communications, and electrical power. Small satellites can take less time to develop, particularly if they use standard, proven, off-the-shelf designs. These enable fast-track science and technology missions.

Some launch vehicle and services companies are providing services specifically to small spacecraft, and a "cottage industry" has grown up around the flight equipment. The advantages of economy of scale are apparent. DARPA has been involved, having developed a launcher for 24 microsatellites at a time. NASA's Ames Research Center in California has a major role in CubeSats. A new initiative using CubeSats for lunar and planetary exploration is emerging

In the taxonomy of smaller payloads, a "small" satellite or mini-sat masses between 100 and 500 kg. A nanosatellite is defined as having a mass between 1 and 10 kg. I can pick that up off the floor and put it on the bench. In 2014, close to 100 nanosats were launched, and 1000 are projected to go to orbit in the next 5 years. In November of 2013, 29 nanosats went to orbit on one launcher from Wallops Flight Facility in Virginia. Just 2 days later, 32 nanosats went to orbit on a Russian booster. In January of 2014, 33 satellites went to the International Space Station on 1 flight.

A picosat has a mass between 0.1 an 1 kg. These are attractive to the do-it-yourself community. A Cubesat, discussed below, is an example of a picosat. Picosats the size of a can of soda can get to orbit for $20,000 or less.

But, they're not the smallest. The Femtosat has a mass between 10 and 100 grams. Some of these are referred to a a single-chip-sat. Several of these went up to the Space Station in 2011, and are attached to an external mount. An interesting mission in March 2014, the Kicksat, a nanosatellite, carried 104 femtosatellites to orbit. These were part of a crowd-funded project from Cornell University. The parts cost of a unit was around $25. Unfortunately, the satellites were not released from the carrier, but the mission was reflown. They were attached to the outside of the ISS for three years, then retrieved and returned to Earth, in working order. Further missions were flown with this architecture.

Cubesat

A cubesat is a small, affordable satellite that can be developed and launched by college, high schools, and even individuals. The specifications were developed by Academia in 1999. The basic structure is a 10 centimeter cube, (volume of 1 liter) weighing less that 1.33 kilograms. This allows a series of these standardized packages to be launched as secondary payloads on other missions. A Cubesat dispenser has been developed, the Poly-PicoSat Orbital Deployer, that holds multiple Cubesats and dispenses them on orbit. They can also be launched from the Space Station, via a custom airlock. ESA, the United States, and Russia provide launch services.

Cubesats can be custom made, but there has been a major industry evolved to supply components, including space computers. It allows for an off-the-shelf implementation, in addition to the custom build. There is quite a bit of synergy between the Amsat folks and Cubesats. NASA supports the Cubesat program, holding design contests providing a free launch to worthy projects. Cubesats are being developed around the world, and several hundred have been launched. Interestingly, they are compatible with every current launch vehicle, and are carried in lieu of ballast (use to adjust the center of mass). A single launcher can carry 30 Cubesats in addition to their primary payload.

The Cubesat STRaND-1 used a Google Nexus smartphone as an onboard computer. Why not? The phone has an accelerometer, a magnetometer, gps, a gyroscope, a barometer, a camera or two, and a radio.

Build costs can be lower than $10,000, with launch costs ranging around $100,000, a most cost-effective price for achieving orbit. The low orbits of the Cubesats insure eventual reentry into the atmosphere, so they do not contribute to the orbital debris problem.

A simple Cubesat controller can be developed from a standard embedded platform such as the Arduino. The lack of radiation hardness can be balanced by the short on-orbit lifetime. The main drivers for a Cubesat flight computer are small size, small power

137

consumption, wide functionality, and flexibility. In addition, a wide temperature range is desirable. The architecture should support a real time operating system, but, in the simplest case, a simple loop program with interrupt support can work.

If you want guaranteed performance with radiation hardened hardware, it will cost more, but quite a few vendors are available. Here are a few examples.

The NanoMind A712D is an onboard computer for Cubesats. It uses as 32-bit ARM cpu, with 2 megabytes of RAM, and 8 megabytes of flash memory. It can also support a MicroSD flash card. It has a Can bus and a I^2C interface. It comes with an extensive software library and real time operating system. Special applications, such as attitude determination and control code are available. It is tolerant to temperatures form -40 to 85 degrees C, but is not completely rad-hard.

The CFC-300 from InnoFlight Inc. of San Diego is another example. It uses the Xilinx Zynq System-on-a-chip architecture. That provides both FPGA capability, and an Arm Cortex A-9 dual core cpu. It has 256 Megabytes of SDRAM, and 32 megabytes of flash. There are multiple synchronous serial interfaces. Daughter cards provide support for SpaceWire, Ethernet, RapidIO, RS-422, and thermistor inputs and heater drive outputs. It can be used with linux or VxWorks.

The Intrepid Cubesat OBC from Tyvak Uses a 400 MHz Atmel processor, and has 128 Mbytes of SDRAM, and 512 Mbytes of flash memory. It draws between 200-300 milliwatts. It includes a command and data handling system, and an onboard electrical power controller. It supports ethernet, RS-232, USB, and the SPI and I^2C interfaces. It includes a JTAG debugging interface. Similar to the Arduino, it supports 3-axis gyros, a 3-axis magnetometer, accelerometers, and a variety of i^2c-interfaced sensors. The Microcontroller is an ARM architecture, with digital signal processing extensions. It has a built-in Image Sensor interface.

COVE is JPL's Xilinx Virtex-5 FPGA-based onboard processor for Cubesats. The FPGA is rad-hard. This high end machine provides sufficient power for onboard data processing, while providing a low power mode for periods where the number crunching is not needed. The FPGA can be reconfigured in flight. It has flown in space several times.

An Australian satellite, Fedsat, uses a Xilinx FPGA with mitigation in the hardware to implement a "self-healing" architecture .It was launched in 2002, and the mission lasted 18 months, before a battery failure brought it to an end.

The Yaliny flight computer is based on the Microsemi Igloo-2 FPGA SOC. It is inherently SEU-immune. There is a soft processor core, implemented withing the FPGA. It has 8 megs of non-volatile, error-correcting memory, and 16 megs of static ram with error-correction. There is the ability to support 1 gigabyte of DDR SDRAM with error correction. It supports the 1553 bus, Ethernet, RS-485 quad pci-express busses, and usb (for debugging). The processor dissipates 2 watts nominally.

The Indian PISAT (PESIT Imaging Satellite) Cubesat uses an Atmel AVR32 32-bit microcontroller as an onboard computer/controller.

The proliferation of low cost and hobbyist grade Flight Computers can only have a positive effect on making the next generation of spacecraft smarter and cheaper.

At NASA and many National Labs, Cubesats have been a game-changer. The cost to develop, build, and test a concept or technology has gone down by orders of magnitude. This precursor technology has not only gone down in price, but the implementation process has been accelerated..

A recent NASA/GSFC Cubesat project, Dellingr, This will be a

6U (12" x 8" x 4") size. It was a one-year project to design, develop, test, and integrate the unit. It will be heading to the International Space Station. It is a Heliophysics payload, carrying an ion/neutral mass spectrometer. The design will be made available as Open Source after the mission is kicked off. It was carried to the ISS by a Dragon resupply flight, and launched from there in November of 2016. It uses the PiSat architecture. The Pi-sat architecture came out of the Flight Software Branch of NASA's Goddard Space Flight Center. It is based on a Cubesat architecture, with a Raspberry Pi as a flight computer. This is a $35, COTS product. It runs linux, and the CFS. It hosts the I2C, SI,, Ethernet, and USB interfaces. It supports the Xbee Mesh network for distributed missions.

Pi-Sat is a project from NASA-GSFC's Flight Software Branch, Code 582. It is intended as a standard low cost satellite and Distributed Mission Test Platform. It incorporates an ARM-based Raspberry Pi processor. The unit runs NASA's Core Flight Software. The computer supports an SD card for storage, and includes wifi and A/D. It is built to the 1U form factor.

Another project was the NSF-funded Firefly mission, launched in November of 2013, and now returning good data on terrestrial Gamma ray flashes, These are interesting phenomena, involving high energy electrons generated by thunderstorms. Firefly uses a Pumpkin flight Motherboard for avionics, based on the Texas Instruments MSP430 chip. That unit is a 16-bit risc microcontroller architecture, The unit is ultra-low power, and mixed signal, supporting analog. It includes a real-time clock, and non-volatile FRAM memory.

The Mars Cube One (MarCo) project from JPL is a set of two Cubesats going to the Red Planet This is a game changer, applying the standardization, modularity, and low cost of cubesats to missions beyond Earth. MarCo is a 6U cubesat design, and will be the first interplanetary cubesat mission. Two cubesats will be launched along with the InSight Mars lander, but will proceed on their own to Mars, where they will conduct a flyby. The units have

140

solar panels for electrical power. Their primary role is as a communications relay.

The Spacecraft Supercomputer Project

This work was supported in part by NASA/Goddard Space Flight Center during 1991.

The goal of the effort was to define architecture with an order of magnitude performance increase over existing spacecraft onboard computing resources. This goal was exceeded, by exploiting scalable parallel architectures.

This study established the feasibility of using off-the-shelf hardware with a known path to full space qualification to address a large set of user processing needs in the area of sensed data sets for Earth Observation, including weather, data. It is no longer feasible to collect large volumes of data that are downlinked over already strained communication channels to large central archive and processing centers based on old-generation computational resources. Now, the choice is between not collecting the data versus generating useful data products close to the data source, and directly downlinking these to users in the field, in addition to and in parallel with the classical data collection and downlink tasks. In many cases, the user's needs are for the receipt of timely data products directly at a remote site. This is not cost effective or even feasible for most current or planned data processing facilities for downlinked sensed data.

One major focus of the High Performance Computing and Communications (HPCC) Program is "in making parallel computing easier to use and scalable..." NASA's role in HPCC is "to accelerate the development and application of high performance computing technologies to meet NASA's science and engineering requirements." This applies to NASA's ground based systems, and also to flight systems, which lag further and further behind. Application of these techniques involve a major effort in system design. The application of new architectures involve the development of new toolsets, software, and paradigms.

This effort focused on extracting information from raw data onboard the spacecraft at the sensor, and established the feasibility of generating data products onboard for direct downlink in a timely manner by using proven hardware, architecture, and algorithms. This approach does not preclude or interfere with the normal collection and archiving of sensed data, but rather works with data tapped off the main stream, and processed in a parallel stream.

Scalable parallel processing techniques are applicable to a large set of spacecraft onboard processing tasks now and in the immediate future. This application will provide the capability to generate and rapidly distribute data products that cannot otherwise be done.

At the time, the Transputer was the best candidate for implementation of this approach. The Transputer is supported by an Ada compiler from Alsys Corp, as well as numerous other languages with parallel extensions.

Although the goal of this study was to define an architecture with an order of magnitude performance increase over existing onboard computing resources, it was shown that several orders of magnitude were feasible. With scalable processor/communication resources, the hardware can be more appropriately matched to the problem domain, while retaining redundancy and reprogrammability.

The early phases of the study identified a series of candidate science payloads or instruments that the Flight Supercomputer can provide services to. The key goal in this phase was to identify a real application that can be implemented without impacting the instrument schedule or mission success, but that would allow the collection of data that would otherwise be lost. An observing class instrument was preferred, as it would provide a large data source.

Requirements for throughput and processing were derived from EOS instruments-class data. These data provided a strawman set of requirements. Numerous applications were found that could benefit from the utilization of parallel processor technology. These all revolved around the onboard generation and direct downlink of data products of a local interest in a timely fashion to end users in the field. In most cases, the timeliness issue precluded the downlink of data to a classical ground processing facility for data product generation, followed by a dissemination, possible by re-uplink and rebroadcast. The complexity of the onboard system interacts with the complexity, and thus cost, of the ground station equipment. This introduced the concepts of data compression/decompression for transmission. The ground receiving station, exclusive of the RF portion, was considered to be a laptop class computer, augmented with a front-end processor, probably based on a complementary scalable parallel processor. This front-end processor would be a custom designed box. In most cases, the ground based application would be cost sensitive.

Space Computer Corp. of California was under contract to DARPA to produce a "Miniaturized, Low-Power Parallel Processor". Their approach has been to use the Transputer as a communication element for Vector co-processors. A prototype system for guided missile applications was delivered in April 1990, and provided a peak processing throughput of 1.3 Gflops. Current efforts focus on micro miniaturization of the technology, using custom designed ASIC's and wafer scale integration.

The applications for the resulting device include sensor image processing, and Synthetic aperture radar (SAR) processing, including image compression tasks. The SCC-100 is a multi-node device, with each node consisting of a Transputer, memory, and Zoran vector signal processor chips. The Zoran chips provide the computational throughput, and the Transputers provide communications and control. Flight units would require the availability of radiation-hard, Military-spec die.

There were potential application of the Flight Supercomputer to data processing requirements derived from Earth observing class spacecraft, including weather satellites.

The Earth Observation Satellite platforms have two direct broadcast channels, 1 supporting a 15 Mbps rate, and the other supporting 100 mbps. The high rate channel is nominally dedicated to 1 instrument, but can provide a backup to the nominal TDRSS high rate link. The EOS platform instrument set was still not completely defined, but a representative set was used to determine if the Transputer link I/O was capable of supporting the collection of data. Processing of data was not considered, since the algorithms were yet undefined.

In all cases, the data input capacity of a single link on the T-800 Transputer was sufficient to handle the average data rate of the instrument. A single link is also sufficient for the peak rate for all but the HIRIS and ITIR. With 4 links, the Transputer can input more than 1 instrument stream continuously, and with external multiplexing, can handle the entire instrument set. Using data from the an early EOS-A instrument set, the instruments' data could be handled by a T-800 Transputer link, with the exception of the HIRIS instrument in peak mode, which outputted data at 160 Mbps. The EOS instrument-derived requirements enveloped the instrument data rates for Earth observing missions in the near future.

Numerous applications exist in the field of Earth Resources mapping, particularly where asynchronous events directly affect human activities, or require timely response. In many cases, the required data product calculation and distribution must be performed at the data source. This implied the capability of onboard processing of sensed data, and direct downlink of the resultant data products, in parallel with the normal data downlink. It was anticipated that several algorithms could reside in onboard memory, and that code could be uplinked rapidly to implement new or modified algorithms in response to unanticipated events.

One potential data product for onboard calculation involved the determination of the perimeter of a forest fire, which is a classic edge-detection problem, given a thermal band image. Direct downlink to the forest fire command center is essential for freshness of the data. The on-site equipment must be small, lightweight, and inexpensive, so that it can be air-dropped and abandoned if necessary.

The data products of interest to the fire management crews on the ground include the flame front location, direction and rate of spread, smoke plume dimension, and hot spot detection. These data were used for personnel and equipment placement and logistics and safety. Observations from U-2 aircraft as well as TIROS series weather satellites have been used to gather relevant data on fires.

Perimeter determination can be accomplished with a LaPlacian operator, which is homogeneous. This is followed by a scaling, a full rectification, and a thresholding operation. This could be done on a single Transputer at the required data rate. The onboard data storage requirements can be minimized by clever organization of the algorithm. Data storage is a premium item that must be minimized for space flight use. Not only were the storage devices expensive, but they consume resources such as size, weight, and onboard power.

Similar to the determination of the perimeter of a fire is the determination of the area of an oil slick on a body of water. In this case we are interested in the perimeter, but can determine the extent of the "blob", and track the shape, location, and drift. Downlink data can be send directly to the site, and used to position containment equipment. The oil slick identification problem maps easily to the problem of determining the extend of the spread of floodwaters.

Another related problem is the timely determination of the location

145

of schools of fish near the continental shelves, using ocean colorimetry. This process is of interest to commercial fishing fleets, and the timeliness of the information is essential, requiring direct downlink to fishing fleets. Cost of the fleet equipment is also an issue.

Similar applications involve specialized operations to image active volcanoes, or to locate and track the eye of severe storms (hurricanes or typhoons). All of these processes are classical image processing applications that can be hosted on one or several Transputers. In most cases the core algorithm is less than 100 lines of code, but is applied across a real time data set. In this case, a systolic pipeline of Transputers may be the ideal topology.

Using successive images of cloud formations, properly registered by visible land mass, the wind vectors may be inferred by cloud motion. This process involves multi-spectral imaging (visible and infrared) from spacecraft such as SMS and GOES. The derived wind information is of importance in global weather pattern understanding and prediction, and is critical to severe storm forecasting. Implementing this process onboard will result in the ability to generate an downlink a data product that is of timely interest.

The required derivation of the data on the ground has been difficult and time-consuming, and has not been available operationally. The algorithms described in the literature postulate quasi-real time implementation on super minicomputers augmented with array processors. This process could also be implemented on arrays of Transputers, and may be the most promising candidate for further implementation.

Revisiting that part of the study that looked at the initial assumption of using the Transputer, we want to see if that was still valid in the light of advancing technology, and product announcements from other vendors, as well as advances from Inmos.

146

A number of alternate computer and connectivity architectures for the flight supercomputer were examined to establish a taxonomy of choices and grade these according to the realities of schedule and availability. Existing and emerging RISC chips such as the MIPS R3000, the Intel i80860 and 960, the Motorola 96000 DSP series, etc. were examined to determine if there was a better alternative to the Transputer chip, before committing to a processor choice. Of concern were availability, vendor support, and software development tools.

No processor could be found that could provide the same level of connectivity as the Transputer, and the Transputer was ahead in terms of Processor-I/O balance. The current unavailability of the unit in space-qualified versions is the only drawback. The closest second candidate was the recently announced Texas Instruments 'C40 Parallel DSP, which is marketed as a DSP, not a general purpose computer. Because TI has a history of developing Military versions of its products, it is worthwhile to continue to track its development. The communication architecture of the C40 used parallel ports with dma engines, and is necessarily distance-limited. Data on the '040 had only recently become available, and it was not possible to completely evaluate it for the purposes of this study. No other architecture could be found that provided the Transputer's inherent communication capability and connectivity, without extensive glue-logic. In fact, many emerging systems use the Transputer as a communication element for fast RISC processors such as Intel's i860 or Motorola 96000 series. The remainder of this section discusses some of the other RISC architectures that were examined.

The R3000 and Intel 80960 family were selected for the 32-bit follow-on to the 1750A architecture for military avionics by the JIAWG. However, neither provide a connectivity solution comparable to the Transputer. The i80386 was being qualified for use on the Space Station Freedom and with the Flight Telerobotic Servicer, but that chip is not designed for multiprocessing.

As with any space application, the usability of parts will lag the commercial state of the art by 3-10 years. Any new architecture suggested must be available in the correct package, and available with the applicable process screening.

Examining the general problem domain for onboard processing of sensed data, the full spectrum from matrix multiplication (compute-bound) to matrix addition (I/O bound) is seen. Compute bound processes can always be speeded up by faster computational components, faster memory, or a smarter architecture. I/O bound problems present a greater challenge. In fact, it is relatively easy to transform a compute bound problem to an I/O bound problem with a parallel processor. One solution, studied at Carnegie-Mellon University, is the systolic processor, which is a matrix of simple, interconnected processor elements with I/O at the boundaries, and a pipelined processing approach to data that is pulsed through the array. In this scheme (since instantiated in the iWarp product, by Intel), multiple use can be made of each data item, and a high throughput can be achieved with modest I/O rate. There is extensive concurrency and modular expandability, and the control and data flow are simple and regular. This technique, easily implemented on Transputers, lends itself well to repetitive operations on large data sets, such as those generated by spaceborne sensors.

Scalable systems, those made up of multiple computational/communication building blocks, have an architecture that is responsive to the problem domain. In such a homogeneous system, the correct amount of processing and I/O can be provided for the initial requirements, with the ability to expand later in a building block fashion to address evolved requirements as well as redundancy or fault tolerance. Developing software for scalable systems is a challenge, mostly in deciding how the software is spread across the computational nodes. This is a solvable problem, based both on good software tools and on programmer experience. Research into these topics, as well as the

148

ability of the system itself to adapt to processing load, is ongoing.

Of course, the applicability of the parallel processor to a given problem set implies that the applicable algorithm can be parallelized, and a solution can be implemented and debugged in a reasonable time. This implies that an efficient programming and debugging environment exist for the selected hardware. This is certainly the case for Transputer-based systems. The major hurdle is conceptual for the systems integrators - the ability to think in parallel paradigms. This comes with hands-on experience.

The HETE-2 Spacecraft flew four T-805 Transputers, in 2011.

References

"Miniaturized, Low-Power Parallel Processor Technology (Advanced Space Technology Program) Status Briefing," 26 March 1991, Space Computer Corp.

Schwalb, Arthur. "The Tiros-N/NOAA A-G Satellite Series," NOAA Tech Memo NESS 95, Aug. 1979.

"Aircraft and Satellite Thermographic Systems for Wildfire Mapping and Assessment," J. A. Brass, V.G. Ambrosia, P.J. Riggan, J.S. Myers, J.C. Arvesen, AIAA paper AIAA-87-0187.

"Automated Mesoscale Winds Derived from Goes Multispectral Imagery," Robert J. Atkinson (GE), Gregory S. Wilson (NASA/MSFC), (unpublished).

"Design Considerations for EOS Direct Broadcast," Charles H. Vermillion (GSFC), Paul H. Chan (SSAI), 1991.

"Data Volumes and Assumptions," EOSDIS Core System Requirement Specification, 14 Sept, 1990, NASA/GSFC.

Things that can go wrong

In and on the way to space.

ESA's Ariane 5 Launch Vehicle

Ariane 5's first test flight (Ariane 5 Flight 501) on 4 June 1996 failed, with the rocket self-destructing 37 seconds after launch because of a malfunction in the control software. A data conversion from 64-bit floating point value to 16-bit signed integer value to be stored in a variable representing horizontal bias caused a processor trap (operand error) because the floating point value was too large to be represented by a 16-bit signed integer. The software was originally written for the Ariane 4 where efficiency considerations (the computer running the software had an 80% maximum workload requirement) led to 4 variables being protected with a handler while 3 others, including the horizontal bias variable, were left unprotected because it was thought that they were "physically limited or that there was a large margin of error". The software, written in Ada, was included in the Ariane 5 through the reuse of an entire Ariane 4 subsystem despite the fact that the particular software containing the bug, which was just a part of the subsystem, was not required by the Ariane 5 because it has a different preparation sequence than the Ariane 4. The incident resulted in a loss of over $500 million.

There was a flight control system failure. A diagnostic code from failed IRS-2 was interpreted as data. IRS-1 had failed earlier. The diagnostic data was sent because of a software error. The software module was only supposed to be used for alignment, not during flight. The diagnostic code was considered a 64-bit floating point number, and converted to a 16-bit signed integer, but the value was too large. This caused the rocket nozzles to steer hard-over to the side, causing the vehicle to veer and crash into a Mangrove swamp.

References

De Dalmau, J. and Gigou J. "Ariane-5: Learning from flight 501 and Preparing for 502, http://esapub.esrin.esa.it/billetin/bullet89/dalma89.html

Lions, Prof, J. L. (Chairman) ARIANE 5 flight 501 Failure, Report by the Inquiry Board, 19 July 1996, http://www.esrin.esa.it/tidc/htdocs/Press/Press96/ariane5rep.html

Jezequel, Jean-Marc and Meyer, Bertrand "Design by Contract: The Lessons of Ariane," IEEE computer, Jan. 1997, vol. 30, n. 2, pp129-130.

"Inquiry Board Traces Ariane 5 Failure to Overflow Error," http://siam.org/siamnews/general/ariance.html
Baber, Robert L. "The Ariane 5 explosion as seen by a software engineer,"

http://www.cs.wits.ac.za/~bob/ariane5.htm

Mars Climate Orbiter

The spacecraft was lost on Mars in September 1999. The system-level requirements did not specify units, so JPL used SI units and the contractor Lockheed Martin used English units. This was not caught in the review process, and led to the loss of the $125 million mission. The spacecraft crashed due to a navigation error.

Architecture, CPU, Memory, I/O

Single RAD6000 cpu, 128 meg ram, 18 meg flash.

Software

VxWorks operating system with flight software developed at

Lockheed Martin Corp.

Sensors, Actuators

Dual 3-axis gyros, star tracker, dual sun sensors, 8 thrusters, 4 reaction wheels.

Root Cause

The primary cause of this discrepancy was human error. Specifically, the flight system software on the Mars Climate Orbiter was written to calculate thruster performance using the metric unit Newtons (N), while the ground crew was entering course correction and thruster data using the Imperial measure Pound-force (lbf). This error has since been known as the *metric mixup* and has been carefully avoided in all missions since by NASA.

"The 'root cause' of the loss of the spacecraft was the failed translation of English units into metric units in a segment of ground-based, navigation-related mission software, as NASA has previously announced," said Arthur Stephenson, chairman of the Mars Climate Orbiter Mission Failure Investigation Board. "The failure review board has identified other significant factors that allowed this error to be born, and then let it linger and propagate to the point where it resulted in a major error in our understanding of the spacecraft's path as it approached Mars."

Mars Rover Pathfinder

The computer in the Mars Rover *Pathfinder* suffered a series of resets while on the Martian surface, but was later recovered.

Architecture, CPU, Memory, I/O: Single RS-6000 cpu, 1553- and VMEbus.

Software: VxWorks, with application code in c.

Sensors, Actuators:Sun sensors, star tracker, radar altimeter, accelerometers, wheel drive.

Root Cause of the problem

Priority inversion in the operating system. Pre-emptive priority thread scheduling was used. The watchdog timer caught the failure of a task to run to completion, and caused the reset. This was a sequence of tasks not exercised during testing. The problem was debugged from Earth, and a software correction uploaded.

The failure was identified by the spacecraft as a failure of one task to complete its execution before the other task started. The reaction to this by the spacecraft was to reset the computer. This reset reinitializes all of the hardware and software. It also terminates the execution of the current ground commanded activities.

The failure was a classic case of priority inversion (The details of how this was discovered and corrected is a fascinating story – see refs.) The higher priority task was blocked by the much lower priority task that was holding a shared resource. The lower priority task had acquired this resource and then been preempted by several of the medium priority tasks. When the higher priority task was activated, to setup the transactions for the next 1553 bus cycle, it detected that the lower priority task had not completed its execution. The resource that caused this problem was a mutual exclusion semaphore used to control access to the list of file descriptors that the select() mechanism was to wait on.

The select mechanism created a mutual exclusion semaphore to protect the "wait list" of file descriptors for those devices which support select. The VxWorks pipe() mechanism is such a device

153

and the IPC mechanism used is based on pipes. The lower priority task had called select, which had called other tasks, which were in the process of giving the mutex semaphore. The lower priority task was preempted and the operation was not completed. Several medium priority tasks ran until the higher priority task was activated. The low priority task attempted to send the newest high priority data via the IPC mechanism which called a write routine. The write routine blocked, taking the mutex semaphore. More of the medium priority tasks ran, still not allowing the high priority task to run, until the low priority task was awakened. At that point, the scheduling task determined that the low priority task had not completed its cycle (a hard deadline in the system) and declared the error that initiated the reset.

NEAR spacecraft

The Near Earth Asteroid Rendezvous – Shoemaker was launched in 1996 to study the asteroid Eros. It was a JHU/APL spacecraft, and a NASA mission. On Monday, 12 February 2001, the NEAR spacecraft touched down on asteroid Eros, after transmitting 69 close-up images of the surface during its final descent.

However, previously, the first of four scheduled rendezvous burns had been attempted on December 20, 1998. The burn sequence was initiated but immediately aborted. The spacecraft subsequently entered safe mode and began tumbling. The spacecraft's thrusters fired thousands of times during the anomaly, which expended 29 kg of propellant reducing the spacecraft's propellant margin to zero. This anomaly almost resulted in the loss of the spacecraft due to lack of solar orientation and subsequent battery drain. Contact between the spacecraft and mission control could not be established for over 24 hours.

Architecture, CPU, Memory, I/O: Three sets of computers, AIU, FC, and tbd. Two each.

Two redundant 1553 standard data buses, two solid state recorders with 1.1 gigabits and 0.67 gigabits respectively.

Software

80,000 lines of guidance and control code.

AIU: 21,000 lines of c code; 10,000 lines of assembly.

FC: 42,000 lines of ADA, 7,000 lines of assembly.

During the Anomaly Board review, seventeen software errors were uncovered. One of these which reported the momentum wheel speed as zero when it was actually at maximum. There were two different archived versions of the software flight load, and no one knew which was onboard the spacecraft.

Sensors, Actuators

Five digital solar attitude detectors, an inertial measurement unit, (IMU) and a star tracker camera. Four reaction wheels for attitude control. Thrusters to dump angular momentum from the reaction wheels, and for rapid slew and propulsive maneuvers. Gyros.

Root Cause of the problem

Startup transient of the main engine exceeded an (incorrect) lateral acceleration safety threshold; leading to engine shutdown. There was a missing command in the burn-abort contingency script.

Clementine Spacecraft

The Clementine spacecraft suffered a catastrophic loss of propellant on May 7, 1994, leading to loss of the primary mission.

Architecture, CPU, Memory, I/O: Dual Command and Telemetry Processors (CTP), MIL-STD-1750A architecture

Sensors, Actuators: Dual star trackers, dual inertial measurement units, attitude control thrusters, reaction wheels.

Root Cause of the problem

The thrusters were erroneously held open for 11 minutes by the flight computer because the thruster protection timer (in software) contained an undetected bug. This depleted the spacecraft propellant.

Soyuz TMA-1 Flight Computer Problem

The new guidance computer of the Russian Soyuz TMA-1 caused an off-course landing in its first use in 2003. This was a concern for the crew of the International Space Station, as the Soyuz TMA-2 was docked to the station as the crew return vehicle, and it had the same computer. The Soyuz is normally controlled to skim the atmosphere to reduce its velocity, using a deceleration of 5g's. The center of gravity of the craft is off center by design, and by rolling the capsule, the tilt, and thus the lift, can be controlled. As in the Apollo days, too steep, you burn up, and too shallow, you skip off the atmosphere and head for space.

The TMS-1 autopilot lost its references, and switched to backup. This simple mode uses a roll maneuver to even out the path, while resulting in a deceleration twice that of the nominal mode, and a very short landing site, compared to the nominal. In this case, the crew returned safely, and were spotted by a rescue team within two hours.

It could have been worse. In 1965, the crew of the Voskhod-2, which had accomplished the first spacewalk. After reentry, when the capsule landed some 386 kilometers off course, the crew had to spend the night in their capsule, due to the danger of bears and wolves in the area. Welcome back to Earth, tasty morsels!

Initially, the Soyuz error was attributed to the American crewman pushing the backup mode activation button, but this was refuted by the crew, A software cause was sought. Software problems of potentially fatal effect had happened in 1988 (with the crew catching the error in time), and again in 1997, where two potentially catastrophic flaws were caught and mitigated by human intervention. One of these would have fired the reentry rockets in the wrong direction. The software has since been corrected. The Soyuz remains the main delivery and return vehicle for the crew of the International Space Station.

Titan Launch Vehicle

The Titan missile was a mainstay of the ballistic missile deterrent system during the Cold War. Housed in hardened underground bunkers with their guidance computers, the missiles were fueled and ready to retaliate against anyone attacking the United States. They were dispersed geographically across the country. Later, the Titan was used as a space launch vehicle for both military and civilian missions.

In 1999, a Titan IV-B with a Centaur upper stage and Milstar satellite left Cape Canaveral Air Station, bound for geosynchronous orbit. After the Centaur second stage with the payload separated from the Titan booster at 9 minutes into the flight, things started to go very wrong. Instability in the roll axis became pitch and yaw axis instability which became uncontrolled tumbling. The Centaur struggled to control these errors, but depleted its available propellant. The Milstar payload ended up in the wrong orbit. The mission was declared a complete loss by the

Secretary of the Air Force, with a cost of around one billion dollars. This was the third straight failure of a Titan mission, and got extensive interest from the Media.

The Accident Investigation Board reach the conclusion that the problem was due to a "failed software development, testing, and quality assurance process." Human error was the cause of an incorrect entry of a value for the roll rate filter constant. This was seen during testing, but not recognized as an error, the consequences were not appreciated, and the error was not corrected. The software development process was shown to allow single point failures. The Independent Verification and Validation (IV&V) Process was not applied to the roll rate filter constant.

References
sunnyday.mit.edu/accidents/titan_1999_rpt.doc

http://www.spacedaily.com/news/titan-99e.html

http://www.youtube.com/watch?v=ZFeZkrRE9wI

Phobos Grunt

In November of 2010, the Russian Space Agency launched an ambitious mission to set a probe down on the small Martian moon Phobos, collect samples, and return them to Earth. There was a failure of the spacecraft propulsion system that stranded the mission in Earth orbit. It reentered the Earth's atmosphere in January 2011. Various causes were postulated for the failure, including interference by U.S. Radar.

The final report from Roscosmos cited software errors, failure of chips in the electronics, possibly due to radiation damage, and the use of non-flight qualified electronics, with inadequate ground testing.

Evidently, identical chips in two assemblies failed nearly simultaneously, so quickly that an error message was not generated. It was possible that the error was recoverable, as the spacecraft entered a safe mode with a proper sun orientation for maximum power. However, the design precluded the reset mode before the spacecraft left its parking orbit. This was major design oversight.

The identified chips that failed were 512k SRAM. The part numbers from the Russian report were checked by NASA's Jet Propulsion Lab, and were found to be among the most radiation susceptible chips they had ever seen. Bad choice. The chips could last in space a few days, and were barely acceptable for non-critical applications, The probably failure cause was single event latch-up, which is sometimes recoverable. In single event latch up, there is a single particle strike that latches up a transistor, preventing it from operating. Usually, if you turn it off and back on again, it will work. A lot of radiation damage to the underlying semiconductor lattice fixes itself after a while, a process called "annealing."

References

Klotz, Irene "Programming Error Doomed Russian Mars Probe," Discovery News, Feb. 7, 2012, news.discovery.com

de Carbonnel, Alissa "Russia races to salvage stranded Mars probe, " Reuters, 2011. www.reuters.com

Amos, Jonathan "Phobos-Grunt mars Probe loses its way just after launch," 9 Nov. 2011, BBC News, www.bbc.co.uk

Oberg, James "Did Bad Memory Chips Down Russia's Mars Probe?," Feb 2012, IEEE Spectrum, IEEE.org.

Friedman, Louis D. "Phobos-Grunt Failure Report Released,"

2/6/2012, www.planetary.org/blogs/guest-blogs/lou-friedman

"Phobos fail: What really happened to Russia's Mars Probe?," Jan 19, 2012, RT.com.

<u>Afterthought</u>

"Man is the best computer we can put aboard a spacecraft... and the only one that can be mass-produced with unskilled labor." attributed to Dr. Wernher von Braun.

Glossary

1553 – Military standard data bus, serial, 1 Mbps.

6u – 6 units in size, where 1u is defined by dimensions.

8051- 8 bit microcontroller from Intel.

AACS – (JPL) Attitude and articulation control system.

ACE – attitude control electronics

Ada – a computer language.

AFB – Air Force Base.

AFMTC - Air Force Missile Test Center

AFRL – Air Force Research Lab

AGC – Automated guidance and control.

AIAA – American Institute of Aeronautics and Astronautics.

AIST – NASA GSFC Advanced Information System Technology .

ALU – arithmetic logic unit.

AmSat – Amateur Satellite. Favored by Ham Radio operators as communication relays.

AOP – Advanced onboard processor (NASA/GSFC).

AP – application programs

AP-101 – Based on IBM S/360 technology, the computers on the Space shuttle

APL – Applied Physics Laboratory, of the Johns Hopkins University.

Apollo – US crewed lunar program.

Arduino – a small, inexpensive microcontroller architecture.

Argon – Soviet flight computer family.

ARPA – (U. S.) Advanced Research Projects Agency

ASIC – application specific integrated circuit

ASIN – Amazon Standard Inventory Number

Async – not synchronized

ATAC – Applied Technologies Advanced Computer

Athena – a missile guidance computer for the Titan-I

ATS-6 – Applications Technology Satellite – 6.

BAE – British Aerospace.

Beowolf – a cluster of commodity computer chips'; multiprocessor.

Bit – binary variable, value of 1 or 0.

Bomarc – early U.S. Cruise missile.

Bootloader – initial program run after power-on or reset. Gets the computer up & going.

BSP – board support package. Customization Software and device drivers.

BURAN – Soviet Space shuttle.

Burroughs Mod 1 – Atlas missile guidance computer.

byte – a collection of 8 bits

cache – temporary storage

CAN - controller area network bus.

C&DH – Command and Data Handling (unit)

cFE – Core Flight Executive.

CFS – Core Flight Software

CME – Coronal Mass Ejection. Solar storm.

CMOS – complementary metal oxide semiconductor.

CPCI – bus, compact PCI.

Coldfire – 32 bit processor from Motorola/Freescale S semiconductor.

Cordic – Co-ordinate Rotation Digital Computer – to calculate hyberbolic and trig functions.

Cots – commercial, off the shelf

CPU – central processing unit

CRC – cyclic redundancy code – error detection and correction mechanism.

CCSDS – Consultive Committee on Space Data Systems.

CubeSat – small inexpensive satellite for colleges, high schools, and individuals.

DARPA – Defense advanced research projects agency.

dc – direct current.

D-cache – data cache.

DDR – dual data rate memory.

DF-224 – flight computer on Space Telescope.

Discrete – single bit signal.

DMA – direct memory access.

DOC – Digital Operations controller, on ATS-6 spacecraft.

DOD – (US) Department of Defense.

Downlink – from space to earth.

162

Dram – dynamic random access memory.

DSP – digital signal processing/processor.

ECC – error correcting code

EDAC – error detecting and correction circuitry.

Eeprom – Electrically erasable programmable read only memory.

EDAC – error detection and correction.

Eniac – very early mainframe computer.

EOS – Earth Observation spacecraft.

ESA – European Space Organization.

Eps – electrical power subsystem

ESRO – European Space Research Organization

ESTO – NASA/GSFC – Earth Science Technology Office.

Ethernet – networking protocol.

ev – electron volt, unit of energy

EVA – extra-vehicular activity

Falcon – launch vehicle from SpaceX

Firewire – IEEE-1394 standard for serial communication.

Flash – non-volatile memory

Flatsat – prototyping and test setup, laid out on a bench for easy access.

FlightLinux – NASA Research Program for Open Source code in space.

FPGA – field programmable gate array.

Fram – ferromagnetic RAM; a non-volatile memory.

FSW – flight software.

Gbyte – 10^9 bytes

Gemini – U. S. crewed 2-person spacecraft.

GeV – billion (10^9) electron volts

Gnu – recursive acronym, gnu is not unix

gpio – general purpose input output

GPS – Global Positioning system – Navigation satellites.

GSFC – Goddard Space Flight Center, Greenbelt, MD

HAL/S – computer language

Hertz – cycles per second

HETE – High energy Transient Explorer (NASA-GSFC).

HPCC – High Performance Computing and Communications.

HPSC – High Performance Spaceflight Computing (NASA)

HST – Hubble Space Telescope
HUD – heads up display.
Icache – Instruction cache
IEEE – Institute of Electrical and Electronic engineers
IEEE-754 – standard for floating point representation and
    calculation.
IEN – Integrated electronics module.
I2C – inter-integrated circuit (serial I/O).
IP – intellectual property; internet protocol.
IP-in-Space – Internet Protocol in Space
ISBN – International Standard Book Number
ISS – International Space Station Freedom.
ITAR – International Trafficking in Arms Regulations (US Dept.
    of State)
IUE – International Ultraviolet Explorer spacecraft
IVHM – Integrated vehicle health magagement
IV&V – Independent validation and verification.
JHU – Johns Hopkins University.
JPL – Jet Propulsion Laboratory
JSC – Johnson Space Center, Houston, Texas.
JWST – James Webb Space Telescope – follow-on to Hubble.
Kbps – kilo (10^3) bits per second.
Kg – kilogram.
kHz – kilo (10^3) hertz
KVA – kilo volts amps – a measure of electrical power
Ku band – 12-18 Ghz radio
Lan – local area network, wired or wireless
Landsat – Earth Observation series of spacecraft.
LaRC – (NASA) Langley Research Center.
Latchup – condition in which a semiconductor device is stuck in
    one state.
Lbf – pounds-force.
LCD – liquid crystal display.
Let- Linear Energy Transfer
LEM – Lunar Excursion Module.
Lidar – optical radar.
Linux – open source operating system

Mbps – mega (106) bits per second.
Mbyte – one million bytes
MER – Mars Exploration Rover
Mercury – early US crewed program.
MEV – million electron volts.
MHz – one million (10^6) Hertz
Minuteman – a solid fueled US ballistic missile.
Microcontroller – monolithic cpu + memory + I/O
Microprocessor – monolithic cpu.
Microsat – satellite with a mass between 10 and 100 kg.
Microsecond – 10^{-6} second
MRAM – magnetorestrictive random access memory.
mSec – Millisecond; (10^{-3}) second.
MIPS – millions instruction per second.
MISC - minimal instruction set computer.
MMS – MultiMission Modular Spacecraft (Nasa-GSFC), also Magnetospheric Multiscale Mission.
MMU – memory management unit; manned maneuvering unit.
MPSoC – Multi Processor System on a chip
Mutex – a software mechanism to provide mutual exclusion between tasks.
Nano – 10^{-9}
NanoSat – small satellite with a mass between 1 and 10 kg.
NASA - National Aeronautics and Space Administration.
NSF – (U.S.) National Science Foundation.
NSSC-1 - NASA Standard Spacecraft Computer – GSFC, 18-bit machine.
NSSC-II - NASA Standard Spacecraft Computer, model 2.
NTRS – NASA Technical Reports Server, www.sti.nasa.gov
OAO – Orbiting Astronautical Observatory
OBC – on board computer
OBP – On Board Processor – NASA/GSFC.
OpAmp – (Linear) operational amplifier; linear gain and isolation stage.
OpCode – encoded computer instruction.
OSS – (NASA) Office of Space Science
Pci – personal computer interface (bus).

Phonesat – small satellite using a cell phone for onboard control and computation.

PI – Principal Investigator

Picosat – small satellite with a mass between 0.1 and 1 kg.

PLL – phase locked loop.

Polaris – U.S. Submarine launched ballistic missile

POSIX – IEEE standard operating system.

PowerPC – 32/64 bit cpu from Motorola/IBM

PPL – preferred parts list (NASA).

Psia – pounds per square inch, absolute.

PSP – Platform Support Package

pwm – pulse width modulation.

Rad – unit of radiation exposure

Rad750 – A radiation hardened IBM PowerPC cpu.

RAID – redundant array of inexpensive disks.

Ram – random access memory

RHBD – rad hard by design.

RHPPC – Rad-Hard Power PC.

RISC – reduced instruction set computer.

RS-232/422/423 – asynchronous and synchronous serial communication standards.

RT – remote terminal.

RTOS – real time operating system.

SDRAM – synchronous dynamic random access memory

SEU – single event upset (radiation induced error)

SIMD – single instruction, multiple data.

SLVGC – Saturn Launch Vehicle Guidance Computer.

SOC – system on a chip.

STAR – self test and repair.

SAA – South Atlantic anomaly. High radiation zone.

Saturn – US heavy lift, manned rocket.

SEU – single event upset.

Simd – single instruction, multiple data

Soc – state of charge; system on a chip.

SoS – silicon on sapphire – an inherently radiation-hard technology

SPARC – risc 32/64 bit computer architecture.

Spi – serial peripheral interface
SSTL – Surrey Space Technology Labs (UK).
SDVF – Software Development and Validation Facility.
SMM – Solar Maximum Mission.
SMP – symmetrical multi-processor
SpaceCube – an advanced FPGA-based flight computer.
SpaceWire – networking and interconnect standard.
Space-X – commercial space company.
Spi – serial peripheral interface
SRAM – static random access memory.
Suitsat – old Russian spacesuit, instrumented and launched from
    the ISS.
SUROM – start-up rom
sync – synchronize, synchronized.
TCP/IP – Transmission Control Protocol/Internet protocol.
TDRSS – Tracking and Data Relay satellite system.
Terrabyte – $10^{housekeep}$
bytes.
T&I – test and integration.
TID – total integrated dose.
Tiros – civilian weather satellite.
Titan – a US ballistic missile, and satellite launcher.
TLE – two line element (position data from tracking.
TMR – triple modular redundancy.
TRMM – Tropical Rainfall Measuring Mission spacecraft.
Transit – early U. S. Navy Navigation satellite system.
Transputer – a 32 bit computer architecture
ttl – transistor-transistor logic integrated circuit.
UART – Universal asynchronous receiver-transmitter.
uM – micro meter
UoSat – a family of small spacecraft from Surrey Space
    Technology Ltd. (UK).
uplink – from ground to space.
USAF – United States Air Force.
Usb – universal serial bus.
V-2 – German World War-2 ballistic missile.
VDC – volts, direct current.

VHDL – very high level design language.
VME – parallel backplane bus architecture.
VxWorks – real time operating systems from Wind River systems.
WiFi – short range digital radio.
Wind River – commercial real time operating system vendor.
Xilinx – manufacturer of programmable logic
Zond – Soviet lunar probe

Bibliography

Alkalai, Leon "An Overview of Flight Computer Technologies for Future NASA Space Exploration Missions," avail: http://www.dlr.de/Portaldata/49/Resources/dokumente/archiv3/100 1.pdf

Alkalai, Leon "NASA's 3D Flight Computer for Space Applications," 1997, NASA/JPL, avail: https://ntrs.nasa.gov/search.jsp?R=20000060822

Alkalai, Leon; Panwar, Ramesh "The Next Generation of Space Flight Computers," 1993, Technology 2003 conference, Anaheim, CA, Dec 7-9, 1993.

Amburn, Brad "NASA Eyes New Generation of Space Computers With More Autonomy' Space News, 06 December 2005.

Asserhall, Stefan "A Processor Transparent On-Board Computer Architecture Using a Radiation Hard Microprocessor, 1998, NRA 98-OSS-10, NASA HQ.

Astrahan, Morton M.; Jacobs, John F. (1983). "History of the Design of the SAGE Computer - The AN/FSQ-7". Annals of the History of Computing (IEEE) 5 (4): 340–349. http://www.livinginternet.com/i/ii_sage.htm

Aukstikalnis, A. J. "Spacecraft computers," 1974, Aeronautics and Astronautics, Vol 12, July-Aug. 1974.

Barnhart, David J., Vladimirova, Tanya, Sweeting, Martin N. "Satellite-on-a-Chip: A Feasibility Study, avail: https://escies.org/download/webDocumentFile?id=1698

Benz, Harry F. Onboard Processor Technology Review,

NASA/LaRC, 1990, paper N90-27152.

Binkley, Jonathan F.; Cheng, Paul G.; Smith, Patrick L; Tosney, William F. "From Data Collection to Lessons Learned, Space Failure information Exploitation at the Aerospace Corporation," 2005, ISHEM 2005, First International Forum on Integrated System Health Engineering and Management in Aerospace, NAPA, CA.

Bitzer, John A. Woerner, Ted A. *A4-Fibel* (English translation) Army Ballistic Missile Agency Redstone Arsenal, Al. 1957 ISBN 1-89-4643-14-3.

Blokdyk, Gerardus *RISC-V a Clear and Concise Reference,* 2018, ISBN-0655348662.

Bollber, M.; Feucht, U.; Frank, H.; Rupp, T. "Flight Experience with the ROSAT Attitude Measurement and Control Subsystem," 1994, Revista Brasileira de Ciencias Mecanicas, Vol 16, pp 77-82.

Boswell, David. "The Software of Space Exploration," 3/30/2006. www.onlamo.com/pub/a/onlamp/2006/0-3/30/Software of space exploration.html

Brown, Thomas K; Donaldson, James A. Fault-Protection Architecture for the Cassini Spacecraft, NASA Tech Brief Vol. 20 No. 8, Item #2, JPL Report NPO-19747, August 1996.

Bucer, Allen W. "Magellan Spacecraft and Memory State Tracking: Lessons Learned, Future Thoughts," 1993, JPL, SpaceOps 1992, Proceedings of the Second International Symposium on Ground Data Systems for Space Mission Operations, pp. 209-214, N94-23832 06-66.

Burghduff, Richard D.; Lewis, Jr., James L. *Man-Machine Interface and Control of the Shuttle Digital Flight System,* NASA Paper N85-16894, in Space Shuttle Technical Conference, NASA-

JSC, June 28-30, 1983.

Byrne, F., Doolittle, G. V., Hockenberger, R. W. "Launch Processing System," Jan. 1976, IBM J. Research & Development.

Casteel, David E. Captain, USAF (ret) "Recollections of the SAGE System" avail: http://ed-thelen.org/comp-hist/CasteelSageRecollections.html

Caudle, John M; Colbert, Donald C. *Flight Control Computer for Saturn Space Vehicle,* NASA/MSFC.

Ceruzzi, Paul E. *Beyond the Limits, Flight Enters the Computer Age,* MIT Press, 1989, ISBN 0-262-03143-4.

Chesnokov, V. V.; Shteinberg, V. I. "Argon Family of computers," Computing in the Soviet Space Program, www.mit.edu/Slava/argon-family.htm

Ciecior, W. et al "A Transputer Based On-Board Data Handling System for Small Satellites," AIAA-93-4467-CP.

Cliff, Rodger A. "The SDP-1 Stored Program Computer," IEEE T. Aerospace and Electronic Systems, Vol. AES-4, no. 6, Nov. 1968.

Cliff, Roger A. "The SDP-3, a Computer designed for Data Systems, of Small Scientific Spacecraft," 1969, NASA X-711-69-43.

Coo, Dennis Advanced Flight Computer, Special Study Final Report, NASA Contractor Report 198165, March 1995, NASA Langley Research Center.

Cooper, A.E. and Chow, W.T. Development of On-board Space Computer Systems, IBM Journal of Research and Development, Volume 20, Number 1, Page 5 (1976).

171

Coon, T. R. and Irby. J. E. "Skylab Attitude Control System," IBM Journal of Research and Development, Volume 20, Number 1, Page 58 (1976).

Cressler, John D. (ed); Mantooth, H. Alan (ed) Extreme Environment Electronics, CRC Press; 1st edition, 2012, ISBN 1439874301.

Cudmore, Alan Pi-Sat: A Low Cost Satellite and Distributed Mission Test Platform, NASA/GSFC, avail: https://ntrs.nasa.gov/search.jsp?R=20150023353

Curiel, Alex da Silva; Davies, Phil; Baker, Adam; Underwood, Dr. Carig; Vladimirova, Dr. Tanya Towards Spacecraft-on-a-Chip, Surrey Satellite Technology Ltd. (presentation), avail: https://escies.org/download/webDocumentFile?id=108.

DeCoursey, R.; Melton, Ryan; Estes, Robert R. Jr. "Sensors, Systems, and Next-Generation Satellites X," Proceedings of the SPIE, Vol. 6361 pp 63611m (2006). avail: https://spie.org/ERS17/conferencedetails/sensors-systems-next-generation-satellites

Deutsch, Dr. Leslie J.; Salvo, Chris "NASA's X2000 Program - an Institutional Approach to Enabling Smaller Spacecraft" Advanced Flight Systems Program, Jet Propulsion Laboratory, California Institute of Technology.

Doyle, Richard et al, "High Performance Spaceflight Computing (HPSC) an Avionics Formulation Task, Study Report," 2012, NASA Office of the Chief Technologist, Game Changing Development Program. Avail: https://gameon.nasa.gov/gcd/files/.../HPSCStudyReportExecutiveSummary102212.pdf

Doyle, Richard Next Generation Space Processor (NGSP) High Performance Spaceflight Computing (HPSC) Next Steps at NASA *and AFRL,* 2013, ASIN-B01D54GONO.

Dumont, Brian, "An Introduction to the Athena Computer," May 16, 1969, Oregon State University, avail: https://archive.org/details/cc-69-09_athenaintro_may69

Eickhoff, Jens *Onboard Computers, Onboard Software and Satellite Operations: An Introduction,* 2011, Springer Aerospace Technology, ISBN-3642251692.

Estaves, G.; Leconte, P.; Vissio, G.; Leyre, X. *SuperComputers for Space Application,* Alcatel Space (presentation), 13 July 2005,

Eyles, Don "Tales from the Lunar Module Guidance Computer," Feb. 6, 2004, AAS-04-064, 27th Annual Guidance and Control Conference.

Fanelli, E.; Hecht, H. "The Fault Tolerant Spaceborne Computer," Computers in Aerospace Conference, 1977, AIAA Paper 77-1490.

Faulkner, A. H.; Gurzi, F.; Hughes, E. L. "Magic, An Advanced Computer for Spaceborne Guidance Systems," A C Spark Plug Division, avail: Internet archive, https://archive.org/details/magic-pdf

Fesq, Lorraine; Dvorak, Dan "NASA's Software Architecture Review Board's (SARB) Findings from the Review of GSFC's "core Flight Executive/Core Flight Software" (cFE/CFS), NASA Software Architecture Review Board , Flight Software Workshop, Nov 7-9, 2012.

Fortescue, Peter and Stark, John *Spacecraft System Engineering,* 2nd ed, Wiley, 1995, ISBN 0-471-95220-6.

Foudriat, E. C., Senn, E. H., Will, R. W., Straeter, T. A., "A Progress Report on a NASA Research Program for Embedded Computer Systems Software," AIAA Paper 79-1956, 1979.

Frank, Laurence J.; Hersman, Christopher B.; Williams, Stephen P.; Conde, Richard F. "A General-Purpose MIL-STD-1750A Spacecraft Computer," 1993, 9ᵗʰ AIAA Computing in Aerospace Conference, AIAA paper 93-4465.

Gaona, John I Jr. "A Radiation-Hardened Computer for Satellite Applications", Sandia Document SAND96-2023C, August 1996, avail: https://digital.library.unt.edu/ark:/67531/metadc667224/

Garman, John R. "The Bug Heard 'Round the World," Oct. 1981, Software Engineering Notes, V 6 n5 p 3-10.

Gasperson, Tina, "FlightLinux blasts off again," Linux.com, July 03, 2007.

Gaudin, Sharon "NASA upgrades Mars Curiosity software… from 350M miles away," Computerworld, article 9230151.

Geer, Dwight A. "System Flight Computer," June 6, 2001, NASA/JPL, Europa Orbiter/X2000 Avionics Industry Briefing.

Gerovitch, Slava "Computing in the Soviet Space Program," http://web.mit/edu/slava/space/introduction.htm

Glenn, Norman P. The new AP101S General-Purpose Computer (GPC) for the Space Shuttle IEEE Proceedings Volume: 75 Page: 308-319, Mar 01, 1987.

Gray, George, "Some Burroughs Transistor Computers (including the Atlas Guidance computer)", Unisys History Newsletter, Volume 3, Number 1, March 1999.

Gray, George, "Sperry Rand Military Computers, 1957-1975," Unisys History Newsletter, Volume 3, Number 4, August 1999.

Green, Chuck; Gulzow, Peter; Johnson, Lyle, Meinzer, Karl;

Miller, James "The Experimental IHU-2 Aboard P3D," Proceedings of the 16th annual AMSAT-NA Space symposium 1998. www.amsat.org/amsat/articles .

Greenemeier, Larry "RS/6000 goes from Deep Blue to the Red Planet," July 11, 1997, www.midrangesystems.com/archive/1997.

Grim, Clifton "IBM FSD 80386 Space Processor DMS Software Architecture Overview", Jan 1988, IBM.

Gruen, Jim, "Linux in Space," 8/24/12, SpaceX, avail: https://events.static.linuxfound.org/images/stories/pdf/lcna_co2012 _gruen.pdf

Hall, Eldon C. *Journey to the Moon: The History of the Apollo Guidance Computer*, 1996, AIAA Press, ISBN 1-56347-185-X.

Harland, David M. and Lorenz, Ralph D. *Space Systems Failures, Disasters and Rescues of Satellites, Rockets and Space Probes*,. 2005, Praxis Publishing, ISBN 0-387-21519-0.

Hartenstein, Raymond G.; Trevathan, Charles E.; Stewart, William N. *The Advanced On-Board Processor, AOP,* NASA/GSFC October, 1971, NASA TM-X-65785.

Hartenstein, R.; Novello, J.;Taylor, T.; Tharpe, M.; Trevathan, C. *On-board Spacecraft Processor*, August 1, 1969, NASA, Document 196900061932 (avail - Nasa Technical Reports Server).

Hartenstein, R.; Taylor, T.; Trevathan, C. E. *An Onboard Processor for OAO-C ,Jan 1, 1969, NASA, Document 19690063747.*(avail - Nasa Technical Reports Server).

Hartenstein, R. *An On-board processor (OBP) for OAO C*, in Significant Accomplishments in Technology Symposium, NASA/Goddard Space flight Center, 1970, NASA-295.

Hecht, H. "Fault-Tolerant Computers for Spacecraft", AIAA J. Spacecraft, V14 N10, Oct. 1977.

Haeussermann, Walter "Description and Performance of the Saturn Launch Vehicle's Navigation, Guidance, and Control System," July 1970, NASA TN D-5869.

Hawkins, Bob *Space Launch System, Exploration Class Capability for Deep Space Exploration*, presented at Flight Software Conference, JHU/APL, 2015. (presentation) avail: http://flightsoftware.jhuapl.edu/files/2015/Day-1/HawkinsJHUFlightSoftware2015102215.pdf

Holcomb, Lee "Overview of NASA's Onboard Computing Technology Program," NASA HQ, 1980.

Holmes-Siedle, A. G. and Adams, L. *Handbook of Radiation Effects*, 2002, Oxford University Press, ISBN 0-19-850733-X.

Hopkins, Albert; Alonso, Ramon; Blair-Smith, Hugh; *Logical Description for the Apollo Guidance Computer (AGC-4)*, unclassified, July 8, 1966, MIT Instrumentation Lab.

Jeans, T. G.; Traynar, C. P. "The Primary UOSAT Spacecraft Computer," Radio and Electronic Engineer, Vol 52, Aug-Sept 1982, pp 385-390.

Jones, Dr. Robert L. and Hodson, Dr. Robert F. "A Roadmap for the Development of Reconfigurable Computing for Space," March 23, 2007, NASA-Langley Research Center.

Judas, Paul A.; Prokop, Lorraine E. "A Historical Compilation of Software Metrics with Applicability to NASA's Orion Spacecraft Flight Software Sizing," in Innovations in Systems and Software Engineering, 2011, 7:161-170, Springer.

Katz, Richard B "Design of Memory Systems for Spaceborne

Computers," November, 2007, FSW Workshop-07.

Kee, W. T. *Programming Manual for the D-37C Computer,* 30 Jan. 1965, Autonetics, Division of North American Rockwell.

Keys, Andrew S. et al, "High Performance, Radiation-Hardened Electronics for Space Environments", International Planetary Probes Workshop-r, 2007, avail: https://nepp.nasa.gov/mafa/talks/MAFA07_17_Keys.pdf

Klesius, Michael "Orion's Brain," Air & Space Magazine, September 2007.

Klotz, Irene "NASA Finds Cause of Voyager 2 Glitch," May 19, 2010, Discovery ZNews, www.discovery.com.

Koczela, Louis J. and Burnett, Gerald J. "Advanced Space Missions and Computer Systems," IEEE T. Aerospace and Electronic Systems, Vol AES-4, No. 3, May 1968, avail: NTRS.

Koczela, Louis J. "Study of Spaceborne Multiprocessing," Phase I, 1970, NASA CR-1446, avail: ResearchGate

Kraft, Jr. Christopher C. "Computers and the Space Program: An Overview," Jan. 1976, IBM J. Research & Development.

Krivonosov, Anatoly; Khartron: Computers for Rocket Guidance Systems (Russian, Ukranian), http://web.mit.edu/slava/space/essays/essay-krivonosov.htm .

Lahti, Doyle; Grisbeck, Gary; Bolton, Phil "ISC (Integrated Spacecraft Computer) Case Study of a Proven, Viable Approach to Using COTS in Spaceborne Computer Systems," General Dynamics Information Systems 14th Annual/USU Conference on Small Satellites. Avail: http://klabs.org/DEI/Processor/PowerPC/scc00-iv-4.pdf

Lesko, J. G. Jr. "Landsat 2 On-board computer," 1975, NASA GSFC, ID 19760059861 A (76A42827), avail: NTRS.

Levenson, Nancy G *An Assessment of Space Shuttle Flight Software Development Process*, 1993, National Academy Press, ISBN 030904880X.

Levenson, Nancy G. *Safeware, System Safety and Computers*, 1995, Addison Wesley, ISBN 0-201-11972-2.

Liu, Chung-Yu, "A Study of Flight-Critical Computer System Recovery from Space Radiation-Induced Errors" IEEE AESS Systems Magazine September 2002, pp. 19-25.

Liu, Yuan-Kwei, "Analysis of the Intel 386 and i486 Microprocessors for the Space Station Freedom Data Management Ssytem," NASA Tech Memorandum 103862, May 1991, avail: NTRS.

Longden, Larry; Thibodeau, Chad;Hillman, Robert; Layton, Phil; Dowd, Michael "Designing A Single Board Computer For Space Using The Most Advanced Processor and Mitigation Technologies," Maxwell Technologies, White Paper. www.maxwell.com

Low, George M., *Apollo Spacecraft*, NASA Manned Spacecraft Center, Houston, TX.

Manning, Robert M. "Low Cost Spacecraft Computers: Oxymoron or Future Trend?," 1993, American Astronautical Society, Advances in Astronautical Sciences, Vol. 81.

Malik, Tariq, "Thinking on Mars: The Brains of NASA's Red Planet Rovers," 2004, www.space.com/businesstechnology

Malinovsky, Boris "Kiev Radio Factory: The First Serially Produced Onboard Computer," (in Russian), www.icfest.kiev.ua/MUSEUM/KRZ.html

178

Marquart, Jane "Flight Software Technology Roadmap, NASA/GSFC, 1998.

Martin, Frederick H. and Battin, Richard H. "Computer-Controlled Steering of the Apollo Spacecraft," J. Spacecraft, Vol 5, n 4, April 1968.

Marshall, Joseph R.; Schimkat, Stuart; Deeds, Matthew W. "Computer systems (Space computing)," December 1999, Aerospace America, pp 30-31.

Martin, Frederick H. and Battin, Richard H. "Computer-Controlled Steering of the Apollo Spacecraft," J. Spacecraft, Vol 5, n 4, April 1968.

McLelland, Mike; Killough, Ronnie "Development of the SC603e VME Based Processor Board," 2000, avail. http://www/ieeecoretech.org/2000/Papers/Hardware.

McMullen, R. F. *History of Air Defense Weapons 1946–1962* (Report), (15 Feb 80), ADC Historical Study No. 14, Historical Division, Office of information, HQ ADC.

Merwarth, A.; Taylor, T.; *Support Software for the Space Electronics Branch On-board Processor, November, 1968, NASA/GSFC, X-562-68-388.*

Messenger, G. C. and Ash, M. S. *The Effects of Radiation on Electronic Systems,* 1992, Van Nostrand Reinhold, ISBN-9401753571.

Moran, Patrick J. "Developing An Open Source Option for NASA Software," NAS Technical Report NAS-03-009, April 21, 2003, NASA/Ames Research Center.
Morrison, Jack C. and Nguyen, Tam "On-board Software for the Mars Pathfinder Microrover," JPL, IAA A-L-0504P.

Nedeau, Joseph; King, Dan; Hun, Ken; Lanza, Denise; Byington, Lester "32-bit Radiation-Hardened Computers for Space," IEEE Aerospace Conference, 1998, ISSN- 1095-323X .

Nguyen, Q., Yuknis, W., Pursley, S., Haghani, N., Albaijes, D. Haddad, O. "A High Performance Command and Data Handling System for NASA's Lunar Reconnaissance Orbiter," AIAA, 7/29/08.

Norman, P. Glenn, "The New AP101S General-Purpose Computer (GPC) for the Space," (IBM Corp.) IEEE, Proceedings Vol 75 Pp. 308-319 Mar, 1987.

Oberg, James "Software Bug sent Soyuz off course," 2003, MSNBC.com. https://spaceflight101.com/meteor-m-2-1/previously-undetected-software-bug-potential-cause-for-soyuz-fregat-failure/

Olsen, Florence, "An IBM rides with Pathfinder," June 2, 1997, Enterprise Computing, Government Computer News.

Patterson, David, Waterman, Andrew *The RISC-V Reader: An Open Architecture Atlas,* 2017, ISBN-0999249118.

Perry, M. A. "The Characteristics and Application of an On-Board Computer for Use on Spacecraft," ESRO Technology Centre, 1974, British Interplanetary Society Journal, Vol. 27, July 1974. pp. 512-520.

Pronobis, Mark T; Selden II, 1Lt. John T. "Radiation-Hardened, 32-bit Processor (RH32) Testability and Fault tolerance, Winter, 1992, RAC Quarterly, Vol 3, Issue 1.

Redmond, Kent C., Smith, Thomas M. *From Whirlwind to MITRE: The R&D Story of The SAGE Air Defense Computer* (History of Computing), MIT Press, October 16, 2000, ISBN-

0262182017.

Reeves, Glenn E. and Ali, Khaled S. The SPIRIT Flash Anomaly Story, FSW-07, JHU/APL.

Rennels, D. A. "Reconfigurable Modular Computer Networks for Spacecraft On-board Processing," 1978, IEEE Computer, Vol. 11, July 1978, pp 49-59.

Robertson, Brent; Placanica, Sam; Morgenstern, Wendy "TRMM On-Orbit Attitude Control System Performance," 1999, American Astronautical Society, 21st Annual AAS Guidance and Control Conference, Parer AAS-99-073

Rocchio, J. J., "Memory Requirements for the Launch Vehicle Digital Computer (LVDC), 1967, Bellcomm, Inc.

http://www.klabs.org/history/history_docs/s5_lvdc/mem_rocchio_67/index.htm

Rooks, John W.; Linderman, Richard "High Performance Space Computing, IEEE Aerospace Conference, 2007, ISSN-1095-323X.

Royal, E. L., *Mariner Venus/Mars 1973 Final Report, Elements (Section 295) of Mission Operations*, November 1974 (JPL Internal Report).

Rubey, Raymond J. Nielsen, William C., Bentley, Laurel; *Flight Computer and Language Processor Study*, NASA CR-1520, March 1970.

Sahu, Kusum EEE-INST-002, *Instructions for EEE Parts Selection, Screening, Qualification, and Derating*, April 2008, NASA/TP-2003-212242.

Schnurr, Richard; Marquart, Jane; Lin, Michael "Standard Spacecraft Interfaces and IP Network Architectures: Prototyping

Activities at the GSFC," NASA/GSFC.

Scott, Barbara "Improving the Onboard Computing Capability of the NASA Multimission Modular Spacecraft,",AIAA Computers in Aerospace Conference, 7th Oct. 1989., avail: https://ntrs.nasa.gov/search.jsp?R=19900023477.

Seagrave, Dorian; Seagrave, Gordon; Godfrey, John; Lin, Michael; "Spacecube: A Reconfigurable Processing Platform For Space," 2008, MAPLD Proceedings.

Shatalov, V. A., Seletkov, S. N., Skrebushevskii, B. S. "Use of Computers in Spacecraft Control System," 1974, Moscow, Izdatel'stvo Mashinostroenie, (in Russian).

Shatalov , Vladimir (Cosmonaut) "Human Being and Computer: A Comparison,"
http://web.mit.edu/slava/space/essays/essay-shatalov.htm

Sklaroff, J. R. *Redundancy Management Technique for Space Shuttle Computers*, IBM J. Research & Development, Jan. 1976.

Smith, Brian *Mongoose ASIC Microcontroller Programming Guide,* NASA Reference Publication 1319, Sept. 1993, NASA/GSFC.

Smith, R. S. "Interpretive Computer simulator for the NASA Standard Spacecraft Computer-2 (NSSC-2)," 1979, NASA Technical Report, NASA-TM-80067

Sorin, Daniel J. and Ozev, Sule, "Fault Tolerant Microprocessors for Space Missions," Duke University.avail: http://citeseerx.ist.psu.edu/viewdoc/download? doi=10.1.1.623.1981&rep=rep1&type=pdf

Speer, David; Jackson, George; Raphael, Dave "Flight Computer Design for the Space Technology 5 (ST-5) Mission, 2002, IEEE

IEEAC paper #254, 0-7803-7231-X.

Speer, John "UARS FOT Significant Events – FW-23," 9 June 1997. "On-Board Computer (OBC) Halt"

Squyres, Steve *Roving Mars: Spirit, Opportunity and the Exploration of the Red Planet,* 2005 Hyperion Books, ISBN 1-4013-0149-5.

Stabler, E. P. and Creveling, C. J. "Spacecraft Computers for Scientific Information Systems," Proc. IEEE Vol. 54 no. 12, December 1966.

Stakem, Patrick H. "A Survey of On-Board Satellite Computers," Orbital Systems, LTD, whitepaper, 1982.

Stakem, Patrick H. "Final Report on the IUE Flight Software Task, under contract NAS5-23830, IUE-MO-77-001, March 9, 1977.

Stakem, Patrick H. "Onboard Computer Study (IUE)," May 18=980, NASA/GSFC.
Stakem, Patrick H. "Operational Experience with Support of a Programmable Spacecraft Onboard Computer," Proc. IEEE Southeastcon, 1977.

Stakem, Patrick H. "Migration of an Image Classification Algorithm to an Onboard Computer for Downlink Data Reduction," AIAA Journal of Aerospace Computing, Information, and Communication , Feb 2004 ,Vol. 1 no. 2 pp 108-111.

Stakem, Patrick H. "Linux in Space", Oct. 2, 2003, invited presentation, Institution of Electrical Engineers, Sheffield Hallam University, Sheffield, UK.

Stakem, Patrick H. "The Applications of Computers and Microprocessors Onboard Spacecraft, NASA/GSFC 1980.

Stakem, Patrick H. "Free Software in Space–the NASA Case," invited paper, Software Livre 2002, May 3, 2002, Porto Allegre, Brazil.

Stakem, Patrick H. "Ground Support Requirements for the Digital Operations Controller onboard Applications Technology Satellite-F," December, 1972, Fairchild Industries.

Stakem, Patrick H. "One Step Forward - Three Steps Backup. Computing in the U.S. Space Program," Byte, Sept. 1981. Reprinted in Military Electronics and Countermeasures, Vol. 7, no.12, and Vol. 8, no.1, Dec 1981 and Jan 1982.

Stakem, Patrick H. "Tracking of Space-Related Flight Computer Systems and the VHSIC Program," BMDSCOM White Paper, 1980.

Stakem, Patrick H. *Microprocessors in Space,* June 22, 2011, PRRB Publishing, ASIN B0057PFJQI.

Stakem, Patrick H. *Apollo's Computers,* 2014, PRRB Publishing, ASIN B00LDT217.

Stevenson, David "Next Generation Embedded Processors Empower Satellite Telemetry and Command Systems, Aeroflex, Colorado Springs, www.aeroflex.com/RadHard

Stewart, W. N.; Hartenstein, R.; Trevathan, C; *Application of an Onboard Processor to the OAO C Spacecraft,* June 1972, NASA/GSFC X-715-72-226, NASA-TM-X-65937.

Stone, John M. Kroesche, Joseph L. Jr. "A Spacecraft Computer for High-Performance Applications," 1992, AIAA Space Programs and Technologies Conference, Huntsville, AL.

Stuhlinger, Ernst, Ordway, II, Frederick I, McCall, Jerry, and Bucher, George C. "From Peenemunde to Outer Space," March 23,

1962, George C. Marshall Space Flight Center.

Styles, F., Taylor, T., Tharpe, M. and Trevathan, C. "A General-Purpose On-Board Processor for Scientific Spacecraft," NASA/GSFC, X-562-67-202, July 1967.

Tantiphanwadi, Dr. Sawat "Spacecraft Computers on the SeaStar Satellite," Orbital Sciences Corporation, 13th AIAA/USU Conference on Small Satellites, SSC99-XII-6.

Taylor, Dan "Curiosity Rolls Ahead on Mars Following Software Upgrade," Information Week, Sept 01, 2012.

Tharpe, M.; Trevathan, C.; Hartenstein, R.; Novello, J. *A General-Purpose On-board Processor for Spacecraft Application,* NASA/GSFC Space Electronics Branch, Oct. 1968, X-562-68-387.

Trevathan, Charles E., Taylor, Thomas D., Hartenstein, Raymond G., Merwarth, Ann C., and Stewart, William N. "Development and Application of NASA's First Standard Spacecraft Computer," CACM V27 n9, Sept 1984, pp. 902-913. avail: http://klabs.org/DEI/Processor/nssc1/p902-trevathan.pdf

Tomayko, James E. *Computers in Spaceflight, The NASA Experience,* NASA contractor Report 182505, 1988. http://history.nasa.gov/computers/contents.html

Tomayko, J. E. "Achieving Reliability: The Evolution of Redundancy in American Manned Spacecraft," Wichita State University, Journal of the British Interplanetary Society, Vol. 38, pp. 545-552, 1985.

Torin, J. M. "The SPOT on board computer," 1979, AIAA Computers in Aerospace Conference, 2nd, A79-54378 24-59.

Van der Velde, W. E., Bentley, G. K., Fagan, J. H., and McDonald, W. T., "Onboard Computer Requirements for Navigation of a

Spinning and Maneuvering Vehicle," J. Spacecraft, Vol. 6 no 12, December 1969.

Vandling, Gilbert C. "Organization of a Microprogrammed Aerospace Computer," February 1975, Computer Design Magazine (AP-101),

Vaugh-Nichols, Stephen J. "Penguins in Space! Asteroid Mining and Linux," Sept. 22, 1013, http://www.zdnet.com.

Walker, Chuck, *Atlas The Ultimate Weapon*, Apogee Books, 2005,ISBN 1-894959-18-3

Wertz, James R. (ed) *Spacecraft Attitude Determination and Control*, section 6.9, Onboard Computers, 1980, Kluwer, ISBN 90-277-1204-2.

Willis, Nathan "LinuxCon: Dragons and Penguins in Space," Sept. 19, 2012, http://lwn.net/Articles/516086/

Wilmot, Jonathan "Implications of Responsive Space on the Flight Software Architecture," 4th Responsive Space Conference, April 24-27, 2006, Los Angeles, Ca.

Wilmot, Jonathan "Overview of Core Flight System (CFS) Implementation of the Goddard Mission Services Evolution Center (GMSEC) Reference Architecture, July 14, 2005, NASA-GSFC.

Wilson, Kevin T. "Analysis of 32-bit Processors for Space System Applications," 1994, Insyte Corporation Technical Report No. 94-2-01.

Wooster, Paul; Boswell, David; Stakem, Patrick; Cowan-Sharp, Jessy "Open Source Software for Small Satellites," SSC07-XII-3, 21st. Annual AIAA/USU, paper SSC07-XII-3, July 2007.

Wooster, Paul; Sharp, Jessy Cowan "The Open Source Space

Software   Community ",   OSCON   2007,
http://conferences.oreillynet.com/cs/os2007/

NASA Publications

Small Spacecraft Technology State of the Art, NASA-Ames,
NASA/tp-2014-216648/REV1, July 2014.

http://spaceflight.nasa.gov/shuttle/reference/shutref/orbiter/avionic
s/dps/gpc.html

NASA Standard Spacecraft Computer (NSSC-II) : Principles of
Operation, 1977, NASA Contractor Report 178826

NASA   Open   Source   Agreement   Version   1.3,
http://opensource.org/licenses/NASA-1.3

AIAA, NASA/GSFC *Aerospace Applications of Microprocessors,*
NASA Conference Publication 2158, NASA/GSFC and AIAA,
Greenbelt, MD, November 3-4, 1980.

Spaceborne Digital Computer Systems, NASA Space Vehicle
Design Criteria (Guidance and Control), NASA SP-8070, March
1971.

New Advanced Computer, Hubble Space Telescope Servicing
Mission 3A, NASA/GSFC, FS-1999-06-009-GSFC.

International Space Station, Command and Data Group, Portable
Onboard Computers, All Expedition Flights, JSC-48529, Dec. 21,
2000, NASA-JSC.

Co-Processor, NASA Facts, NASA/GSFC, NF-193, June 1993
(DF-224 and 80386 upgrade).

Flight   Software   Product   Plan,   Flight   Software   Branch,
NASA/GSFC – code 582, Template 6.0 – 0116/04 document 582-

2000-007.

Rubey, Raymond J., Nielsen, William C., Bentley, Laurel; *Flight Computer and Language Processor Study*, NASA CR-1520, March 1970.

Fault-Protection Architecture for the Cassini Spacecraft, NASA Tech Brief Vol. 20 No. 8, Item #2, JPL Report NPO-19747, August 1996.

X2000 Systems and Technologies for Missions to the Outer Planets. Jet Propulsion Laboratory, Pasadena, CA 49th International Astronautical Congress Sept 28-0ct 2,1998/Melbourne, Australia, IAA-98-IAA.4.1.02.

Core Flight Software System, NASA/GSFC, FS-2009-10-101-GSFC (TT#21).

"NASA Efforts to Develop and Deploy Advanced Spacecraft Computers," March 31, 1989, GAO/IMTEC-89-17.

NASA, "Driving Down Mission costs, New Flight Software Package Delivered to Lunar Mission, http://gsfctechnology.gsfc.nasa.gov/MissionCost.html

NASA, *Radiation Resistant Computers*, Nov. 18, 2005, http://science.nasa.goc/headlines.

Computers Take Flight: A History of NASA's Pioneering Digital Fly-By-Wire Project – Apollo and Shuttle Computers, Airplanes, Software and Reliability, ASIN B006PV3WJY.

Standard Telemetry and Command Components Standard Interface for Computer (STINT) Requirements Document (Preliminary), NASA Goddard Space Flight Center (November 1975). GSFCS-700-51.

Other Sources

Journal of Aerospace Computing, Information, and Communication, AIAA, ISSN 1940-3151.

Flight Software Workshops (FSW), this series of Workshops has been held since 2007, generally alternating between the East Coast and the West Coast. The Aerospace Corporation, NASA Jet Propulsion Laboratory, The Johns Hopkins University Applied Physics Laboratory and Southwest Research Institute are the Founding Sponsors of the Flight Software Workshops. http://flightsoftware.jhuapl.edu/index.html

NASA Office of Logic Design, Spaceborne Processor and Avionics Papers, http://www.solarstorms.org/SEUcomputers.html

wikipedia, various.

https://en.wikichip.org/wiki/microprocessors_used_in_spacecrafts

The Soviets spaceship's on-board computers. http://www.buran-energia.com/bourane-buran/bourane-consti-ordinateur computer.php

The Apollo Flight Journal, The Apollo On-board Computers, http://history.nasa.gov/afj/compessay.htm

Saturn V Launch Vehicle Digital Computer,2 volumes, IBM, 30 Nov 1964.

Kiev Radio Factory: The First Serially Produced Onboard Computer. http://www.icfcst.kiev.ua/MUSEUM/KRZ.html

Shatalov , Vladimir (Cosmonaut) "Human Being and Computer: A

Comparison,"
http://web.mit.edu/slava/space/essays/essay-shatalov.htm

Onboard Processing,
http://www.wtec.org/loyola/satcom2/03_04.htm

Apollo Guidance Computer emulator,
http://www.ibiblio.org/apollo/index.html

Forum on Risks to the Public in Computers and Related Systems.
http://catless.newcastle.ac.uk/php/risks/search.php?
query=spacecraft+computer

(IBM) Space Flight Chronology .
http://www-
03.ibm.com/ibm/history/exhibits/space/space_chronology2.html

The Clementine Spacecraft, Onboard computer failure.
http://astrogeology.usgs.gov/Projects/Clementine/nasaclem/spacecr
aft/spacecraft.html

A High Assurance On-board Computer System for Spacecraft Use
Proceedings of the IEICE General Conference Vol.2000 No.1,
http://ci.nii.ac.jp/naid/110003236919/en/

Gemini Inertia Guidance System,
http://www.astronautix.com/craft/gemystem.htm

Mars Global surveyor, 2007.
http://www.planetary.org/news/2007/0413_Human_and_Spacecraf
t_Errors_Together.html

Argon Onboard Digital Computhttp://milparade.udm.ru/22/78.html

Spaceborne Processor and Avionics Papers, NASA Office of Logic
Design, http://www.solarstorms.org/SEUcomputers.html

"Columbia, other Shuttles have history of computer glitches," Computerworld, Feb. 3, 2003. http://www.computerworld.com/governmenttopics/government/sto ry/0,10801,78135,00.html

Report Reveals Likely Causes of Mars Spacecraft Loss, April 13, 2007. http://www.nasa.gov/mission_pages/mgs/mgs-20070413.html

Kosmonavtka, ISS computers, including laptops. http://suzymchale.com/kosmonavtka/isscomp.html

"The Software Behind the Mars Phoenix Lander," July 9, 2008. http://news.oreilly.com/2008/07/the-software-behind-the-mars-p.html

"Computer sleuths try to crack Pioneer anomaly," March 2, 2007, NewScientist.com

Apollo Guidance Computers, http://klabs.org/DEI/Processor/apollo/index.htm .

Shuttle Avionics, http://klabs.org/DEI/Processor/shuttle/index.htm .

Space Station Avionics, http://klabs.org/DEI/Processor/iss/index.htm .

http://www.silogic.com/Athena/Unisys%20History%20Newsletter %20Aug%201999%20v3n4.htm .

NSSC-1 Onboard Flexibility for Space Missions, IBM Federal Systems Division, Feb. 1978, 78-67K-001.

NSSC-II operating system, design requirements, specification, Intermetrics, Inc. 1979, NASA contractor Report 161396, avail: https://ntrs.nasa.gov/search.jsp?R=198000011579

HP-65 in Space with Apollo-Soyuz,
http://www.hpmuseum.org/adverts/sa65spc.htm

Lockheed Martin, *Software User's Guide for the RAD6000 Processor,*Jan. 20, 2004, Document #204A496.

Missile Systems, Dept. of Missile and Space Training, Missile Launch/Missile Officer (LGM-25), May 1967, Student Study Guide, OBR1821F/3121F-V-1 thru 4, Sheppard Technical Training Center.

Triana Flight Processor User's Guide, Lockheed Martin 233A039, 5/12/1999. (R/6000)

Integrated Spacecraft Controller (ISC) General Dynamics, 2001, avail: www.gd-is.com/sections/products/process/isc1.html

RHPPC single card computer, Honeywell Space Systems, Sept. 1999.

"Air Force Research Laboratory and Honeywell Space systems to develop Space Microprocessor chipset," Press Release, 1/3/2000, 2:51 pm, www.honeywell.com.

RAD750 Radiation Hardened Microprocessor, BAE Systems, 2001, www.baesystems-iews.com/space

RAD750 Microprocessor Datasheet, preliminary Version 1.1, 8/1/00, Lockheed Martin Space Electronics & Communication Systems. (unpublished)

"Sandia Labs to develop custom radiation-hardened Pentium processor for space and defense needs," Press Release, Dec. 8, 1998, Sandia Labs, www.sandia.gov/media

ERC32 datasheets and application notes, ESA, Spacecraft Control and Data Systems Division, www.estec.esa.nl.wsmwww/erc32.

Phillips Laboratory Microprocessor Helps Explore Martian Surface, 1997, http://quark.plk.af.mil/news/archive/97SeptNL/Rad6000.html.

Rad Hard Thor Microprocessor, Saab Ericsson Space, www.saabericssonspace.com/thor

Maxwell Technologies, SCS750 Super Computer for Space, doc 1004741, Rev. 7, www.maxwell.com.

NASA standard spacecraft computer (NSSC-II) : Principles of Operation. NASA Contractor Report 178826.

http://bitsavers.trailing-edge.com/pdf/ibm/sage/SAGE_BOMARC_Defense_System_1958.pdf

http://www.computermuseum.li/Testpage/IBM-SAGE-computer.htm

http://bitsavers.informatik.uni-stuttgart.de/pdf/ibm/sage/22-00001_Central_Computer_System_Preliminary_Sep55.pdf

http://www.mitre.org/about/photo_archives/sage_photo.html

Grimwood, James M.; Stowd, Frances HISTORY OF THE JUPITER MISSILE SYSTEM, 1962, U. S. Army Ordnance Missile Command, http://heroicrelics.org/info/jupiter/jupiter-hist.html

"An Assessment of Space Shuttle Software Development Processes, Committee for Review of Oversight Mechanisms for Space Shuttle Flight Software Processes,"1993, National Academies Press, ISBN-030904880X.

http://www.esa.int/Our_Activities/Space_Engineering_Technology

/Onboard_Computer_and_Data_Handling/Microprocessors.

http://www.jpl.nasa.gov/cubesat/

"One small step toward Mars: One giant leap for supercomputing," August 2017. avail: https://news.hpe.com/one-small-step-toward-mars-one-giant-leap-for-supercomputing/

"The space station gets a new supercomputer," 2017, avail: https:www.hpe.com/us/en/home.html

"Spaceborne Linux Supercomputer Starts Running In Space, Achieves 1 Teraflop Speed," avail: https://fossbytes.com/spaceborne/

High Performance Commercial Off-The-Shelf (COTS) Computer System on the ISS (Spaceborne Computer – 9-20-17. avail: https://www.nasa.gov/mission_pages/station/research/experiments/2304.html

Voyager -https://www.allaboutcircuits.com/news/voyager-mission-anniversary-computers-command-data-attitude-control/

Flight Avionics Hardware Roadmap, NASA TM-2013-217986/REV1, Jan 2014, avail NASA/NTRS, https://ntrs.nasa.gov/archive/nasa/casi.ntrs.nasa.gov/20140011142.pdf .

https://riscv.org

RISC Software Ecosystem, avail - https://riscv.org/wp-content/uploads/2015/02/riscv-software-toolchain-tutorial-hpca2015.pdf

https://wiki.debian.org/RISC-V

http://pluto.jhuapl.edu/Mission/Spacecraft.php#Systems-and-Components.

http://pluto.jhuapl.edu/Mission/Spacecraft.php#Systems-and-Components.

First Russian MIPS-Compatible Microprocessor, avail: http://dailyrumors.blogspot.com/2007/12/first-russian-mips-compatible.html

Microsemi, Mitigation of Radiation Effects in RTG4 Radiation-Tolerant Flash FPGAs WP0191 White Paper. Avail: https://www.microsemi.com/document-portal/doc_view/135027-wp0191-mitigation-of-radiation-effects-in-rtg4-radiation-tolerant-fpgas-white-paper

If you enjoyed this book, you might also be interested in some of these.

Stakem, Patrick H. *16-bit Microprocessors, History and Architecture*, 2013 PRRB Publishing, ISBN-1520210922.

Stakem, Patrick H. *4- and 8-bit Microprocessors, Architecture and History*, 2013, PRRB Publishing, ISBN-152021572X,

Stakem, Patrick H. *Apollo's Computers*, 2014, PRRB Publishing, ISBN-1520215800.

Stakem, Patrick H. *The Architecture and Applications of the ARM Microprocessors*, 2013, PRRB Publishing, ISBN-1520215843.

Stakem, Patrick H. *Earth Rovers: for Exploration and Environmental Monitoring*, 2014, PRRB Publishing, ISBN-152021586X.

Stakem, Patrick H. *Embedded Computer Systems, Volume 1, Introduction and Architecture*, 2013, PRRB Publishing, ISBN-1520215959.

Stakem, Patrick H. *The History of Spacecraft Computers from the V-2 to the Space Station*, 2013, PRRB Publishing, ISBN-1520216181.

Stakem, Patrick H. *Floating Point Computation*, 2013, PRRB Publishing, ISBN-152021619X.

Stakem, Patrick H. *Architecture of Massively Parallel Microprocessor Systems*, 2011, PRRB Publishing, ISBN-1520250061.

Stakem, Patrick H. *Multicore Computer Architecture,* 2014, PRRB Publishing, ISBN-1520241372.

Stakem, Patrick H. *Personal Robots,* 2014, PRRB Publishing, ISBN-1520216254.

Stakem, Patrick H. *RISC Microprocessors, History and Overview,* 2013, PRRB Publishing, ISBN-1520216289.

Stakem, Patrick H. *Robots and Telerobots in Space Applications,* 2011, PRRB Publishing, ISBN-1520210361.

Stakem, Patrick H. *The Saturn Rocket and the Pegasus Missions, 1965,* 2013, PRRB Publishing, ISBN-1520209916.

Stakem, Patrick H. *Visiting the NASA Centers, and Locations of Historic Rockets & Spacecraft,* 2017, PRRB Publishing, ISBN-1549651205.

Stakem, Patrick H. *Microprocessors in Space,* 2011, PRRB Publishing, ISBN-1520216343.

Stakem, Patrick H. Computer *Virtualization and the Cloud,* 2013, PRRB Publishing, ISBN-152021636X.

Stakem, Patrick H. *What's the Worst That Could Happen? Bad Assumptions, Ignorance, Failures and Screw-ups in Engineering Projects, 2014,* PRRB Publishing, ISBN-1520207166.

Stakem, Patrick H. *Computer Architecture & Programming of the Intel x86 Family, 2013,* PRRB Publishing, ISBN-1520263724.

Stakem, Patrick H. *The Hardware and Software Architecture of the Transputer,* 2011, PRRB Publishing, ISBN-152020681X.

Stakem, Patrick H. *Mainframes, Computing on Big Iron,* 2015,

PRRB Publishing, ISBN- 1520216459.

Stakem, Patrick H. *Spacecraft Control Centers*, 2015, PRRB Publishing, ISBN-1520200617.

Stakem, Patrick H. *Embedded in Space,* 2015, PRRB Publishing, ISBN-1520215916.

Stakem, Patrick H. *A Practitioner's Guide to RISC Microprocessor Architecture*, Wiley-Interscience, 1996, ISBN-0471130184.

Stakem, Patrick H. *Cubesat Engineering*, PRRB Publishing, 2017, ISBN-1520754019.

Stakem, Patrick H. *Cubesat Operations*, PRRB Publishing, 2017, ISBN-152076717X.

Stakem, Patrick H. *Interplanetary Cubesats*, PRRB Publishing, 2017, ISBN-1520766173 .

Stakem, Patrick H. Cubesat Constellations, Clusters, and Swarms, Stakem, PRRB Publishing, 2017, ISBN-1520767544.

Stakem, Patrick H. *Graphics Processing Units, an overview*, 2017, PRRB Publishing, ISBN-1520879695.

Stakem, Patrick H. *Intel Embedded and the Arduino-101, 2017,* PRRB Publishing, ISBN-1520879296.

Stakem, Patrick H. *Orbital Debris, the problem and the mitigation*, 2018, PRRB Publishing, ISBN-*1980466483*.

Stakem, Patrick H. *Manufacturing in Space*, 2018, PRRB Publishing, ISBN-1977076041.

Stakem, Patrick H. *NASA's Ships and Planes*, 2018, PRRB

Publishing, ISBN-1977076823.

Stakem, Patrick H. *Space Tourism*, 2018, PRRB Publishing, ISBN-1977073506.

Stakem, Patrick H. *STEM – Data Storage and Communications*, 2018, PRRB Publishing, ISBN-1977073115.

Stakem, Patrick H. *In-Space Robotic Repair and Servicing*, 2018, PRRB Publishing, ISBN-1980478236.

Stakem, Patrick H. *Introducing Weather in the pre-K to 12 Curricula, A Resource Guide for Educators*, 2017, PRRB Publishing, ISBN-1980638241.

Stakem, Patrick H. *Introducing Astronomy in the pre-K to 12 Curricula, A Resource Guide for Educators*, 2017, PRRB Publishing, ISBN-198104065X.
Also available in a Brazilian Portuguese edition, ISBN-1983106127.

Stakem, Patrick H. *Deep Space Gateways, the Moon and Beyond*, 2017, PRRB Publishing, ISBN-1973465701.

Stakem, Patrick H. *Exploration of the Gas Giants, Space Missions to Jupiter, Saturn, Uranus, and Neptune*, PRRB Publishing, 2018, ISBN-9781717814500.

Stakem, Patrick H. *Crewed Spacecraft*, 2017, PRRB Publishing, ISBN-1549992406.

Stakem, Patrick H. *Rocketplanes to Space*, 2017, PRRB Publishing, ISBN-1549992589.

Stakem, Patrick H. *Crewed Space Stations,* 2017, PRRB Publishing, ISBN-1549992228.

Stakem, Patrick H. *Enviro-bots for STEM: Using Robotics in the pre-K to 12 Curricula, A Resource Guide for Educators,* 2017, PRRB Publishing, ISBN-1549656619.

Stakem, Patrick H. *STEM-Sat, Using Cubesats in the pre-K to 12 Curricula, A Resource Guide for Educators,* 2017, ISBN-1549656376.

Stakem, Patrick H. *Lunar Orbital Platform-Gateway,* 2018, PRRB Publishing, ISBN-1980498628.

Stakem, Patrick H. *Embedded GPU's,* 2018, PRRB Publishing, ISBN- 1980476497.

Stakem, Patrick H. *Mobile Cloud Robotics,* 2018, PRRB Publishing, ISBN- 1980488088.

Stakem, Patrick H. *Extreme Environment Embedded Systems,* 2017, PRRB Publishing, ISBN-1520215967.

Stakem, Patrick H. *What's the Worst, Volume-2,* 2018, ISBN-1981005579.

Stakem, Patrick H., *Spaceports,* 2018, ISBN-1981022287.

Stakem, Patrick H., *Space Launch Vehicles,* 2018, ISBN-1983071773.

Stakem, Patrick H. *Mars,* 2018, ISBN-1983116902.

Stakem, Patrick H. *X-86, 40th Anniversary ed,* 2018, ISBN-1983189405.

Stakem, Patrick H. *Lunar Orbital Platform-Gateway,* 2018, PRRB Publishing, ISBN-1980498628.

Stakem, Patrick H. *Space Weather*, 2018, ISBN-1723904023.

Stakem, Patrick H. *STEM-Engineering Process*, 2017, ISBN-1983196517.

Stakem, Patrick H. *Space Telescopes,* 2018, PRRB Publishing, ISBN-1728728568.

Stakem, Patrick H. *Exoplanets*, 2018, PRRB Publishing, ISBN-9781731385055.

Stakem, Patrick H. *Planetary Defense*, 2018, PRRB Publishing, ISBN-9781731001207.

Patrick H. Stakem *Exploration of the Asteroid Belt*, 2018, PRRB Publishing, ISBN-1731049846.

Patrick H. Stakem *Terraforming*, 2018, PRRB Publishing, ISBN-1790308100.

Patrick H. Stakem, *Martian Railroad,* 2019, PRRB Publishing, ISBN-1794488243.

Patrick H. Stakem, *Exoplanets,* 2019, PRRB Publishing, ISBN-1731385056.

Patrick H. Stakem, *Exploiting the Moon,* 2019, PRRB Publishing, ISBN-1091057850.

Patrick H. Stakem, *RISC-V, an Open Source Solution for Space Flight Computers,* 2019, PRRB Publishing, ISBN-1796434388.

Patrick H. Stakem, *Arm in Space*, 2019, PRRB Publishing, ISBN-9781099789137.

Patrick H. Stakem, *Extraterrestrial Life*, 2019, PRRB Publishing, ISBN-978-1072072188.

Patrick H. Stakem, *Space Command*, 2019, PRRB Publishing, ISBN-978-1693005398.